CalcLabs
with Derive®

for Stewart's
MULTIVARIABLE
CALCULUS
Concepts AND Contexts

Jeff Morgan
Texas A & M University

Selwyn Hollis
Armstrong Atlantic State University

BROOKS/COLE PUBLISHING COMPANY
I(T)P® *An International Thomson Publishing Company*

Pacific Grove • Albany • Belmont • Bonn • Boston • Cincinnati • Detroit
Johannesburg • London • Madrid • Melbourne • Mexico City • New York
Paris • Singapore • Tokyo • Toronto • Washington

GWO A GARY W. OSTEDT BOOK

Project Development Editor: *Beth Wilbur*
Marketing Team: *Caroline Croley, Christine Davis*
Field Editor: *Ragu Raghavan*
Editorial Associate: *Carol Benedict*

Production Coordinator: *Dorothy Bell*
Cover Design: *Vernon T. Boes*
Cover Photograph: *David Bruce Johnson, Ian Sabel*
Printing and Binding: *Webcom Limited*

COPYRIGHT © 1998 by Brooks/Cole Publishing Company
A division of International Thomson Publishing Inc.
I(T)P The ITP logo is a registered trademark used herein under license.
Derive is a registered trademark of Soft Warehouse, Inc.

For more information, contact:

BROOKS/COLE PUBLISHING COMPANY
511 Forest Lodge Road
Pacific Grove, CA 93950
USA

International Thomson Editores
Seneca 53
Col. Polanco
11560 México, D. F., México

International Thomson Publishing Europe
Berkshire House 168-173
High Holborn
London WC1V 7AA
England

International Thomson Publishing GmbH
Königswinterer Strasse 418
53227 Bonn
Germany

Thomas Nelson Australia
102 Dodds Street
South Melbourne, 3205
Victoria, Australia

International Thomson Publishing Asia
60 Albert Street
#15-01 Albert Complex
Singapore 189969

Nelson Canada
1120 Birchmount Road
Scarborough, Ontario
Canada M1K 5G4

International Thomson Publishing Japan
Hirakawacho Kyowa Building, 3F
2-2-1 Hirakawacho
Chiyoda-ku, Tokyo 102
Japan

All rights reserved. No part of this work may be reproduced, stored in a retrieval system, or transcribed, in any form or by any means—electronic, mechanical, photocopying, recording, or otherwise—without the prior written permission of the publisher, Brooks/Cole Publishing Company, Pacific Grove, California 93950.

Printed in Canada

10 9 8 7 6 5 4 3 2 1

ISBN 0-534-35740-7

Contents

	Introduction	v
1	**Getting Started**	**1**
	1.1 DERIVE as a Calculator	1
	1.2 Assigning Variables	5
	1.3 Algebra Commands	9
	1.4 Plots	11
2	**Vectors**	**19**
	2.1 Vector Basics	19
	2.2 Length and Arithmetic Operations	23
	2.3 Dot Products and Projections	24
	2.4 The Cross Product	28
	2.5 Equations of Lines	29
	2.6 Equations of Planes	32
3	**Surfaces**	**37**
	3.1 Graphs of Functions of Two Variables	37
	3.2 Traces and Contours	40
4	**Vector-valued Functions**	**47**
	4.1 Space Curves	47
	4.2 Derivatives	52
	4.3 Arc Length and Curvature	56
	4.4 Velocity and Acceleration	65
	4.5 Parametric Surfaces	67
5	**Multivariate Functions**	**71**
	5.1 Limits and Continuity	71

5.2 Partial Derivatives 76
5.3 The Tangent Plane and Linear Approximation 79
5.4 Directional Derivatives and the Gradient 82
5.5 Optimization and Lagrange Multipliers 84

6 Multiple Integrals 95
6.1 Double Integrals 95
6.2 Polar Coordinates 101
6.3 Applications 104
6.4 Triple Integrals 108
6.5 Cylindrical and Spherical Coordinates 113
6.6 Change of Variables 116

7 Vector Calculus 127
7.1 Vector Fields 127
7.2 Line Integrals 132
7.3 Green's Theorem 135
7.4 Surface Integrals 137
7.5 Stokes' Theorem 142
7.6 The Divergence Theorem 143

8 Projects 147
8.1 Osculating Circles 147
8.2 Centers of Circles of Curvature 149
8.3 Coriolis Acceleration 150
8.4 An Ant on a Helix 152
8.5 3D Graphics 153
8.6 Least Squares and Curve Fitting 155
8.7 Classifying Critical Points 157
8.8 Optimization 161
8.9 The Gradient Descent Method 164
8.10 Newton's Method 167
8.11 Balancing a Region 169
8.12 Velocity Fields and Steady Flow 170
8.13 Incompressible Potential Flow 171
8.14 Reflecting Points 173
8.15 Volumes, Surfaces and Tangent Planes 174

Index 176

Introduction

DERIVE is a powerful computer algebra system which can free students from the drudgery of computations, and allow them to concentrate on problem solving. It is also a user-friendly computer algebra system. As such, it seems like a natural choice for students and teachers who want to enhance their calculus experience, and avoid the steep learning curve associated with other computer algebra systems.

This manual was written as a supplement to the first edition of **Multivariable Calculus: Concepts and Contexts**, by James Stewart. The primary goal of this manual is to show how DERIVE can help students learn and use multivariable calculus. To this end, Chapters 2 through 7 contain a number of examples and exercises which parallel the material in Stewart's text. In addition, Chapter 8 contains 15 projects which reinforce and enhance many of the central topics in multivariable calculus.

There is no technology prerequisite for using this manual. DERIVE is so easy to use that almost any student can become comfortable with the interface in 30 minutes. A basic overview of the algebraic and graphic concepts of DERIVE can be found in Chapter 1. Advanced DERIVE users can move immediately to Chapter 2.

As a word of warning to students, DERIVE alone will not solve calculus problems. "Thinking" is required for problem set-up. DERIVE can only be used to help with the computations and graphics that are necessary to obtain final answers.

All of the DERIVE syntax and examples presented in this manual use the Windows version of DERIVE. This manual assumes that the reader is using a PC Windows platform. Non-windows users can still use this manual but they will have to make some modifications in syntax.

Historical Information

In January of 1997, Gary Ostedt asked Jeff Morgan to serve as an Editor for a series of technology manuals designed for use with James Stewart's **Calculus: Concepts and Contexts, Single Variable**. The initial manuals in the series were adaptations of the *Calclabs with Maple* and the *Calclabs with Mathematica* manuals published by Brooks/Cole. Subsequent manuals were developed for the DERIVE computer algebra system, and for the TI-82/83, TI-85/86 and TI-92 calculators. Later that year, DERIVE, Maple, Mathematica and the TI-92 were selected as platforms for supplements to James Stewart's **Multivariable Calculus: Concepts and Contexts**.

All of the author royalties on sales of this manual, have been donated to an undergraduate scholarship fund for Mathematics majors at Texas A&M University.

*Please send all comments and corrections to
Jeff Morgan, Department of Mathematics, Texas A&M University,
jmorgan@math.tamu.edu.*

Chapter 1

Getting Started

This chapter introduces some of the basic **DERIVE** commands associated with assigning variables and creating plots. Experienced **DERIVE** users should proceed to Chapter 2.

1.1 DERIVE as a Calculator

DERIVE input lines are created by *authoring* expressions. Expressions can be authored two different ways. The most obvious way to author an expression is to click on the **Author** menu item and then click on **Expression**. The menu window is shown below.

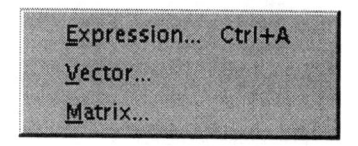

```
Expression...  Ctrl+A
Vector...
Matrix...
```

The other way to author an expression is to press [Ctrl]+A. Which ever method is used, the **Author Expression** window will appear.

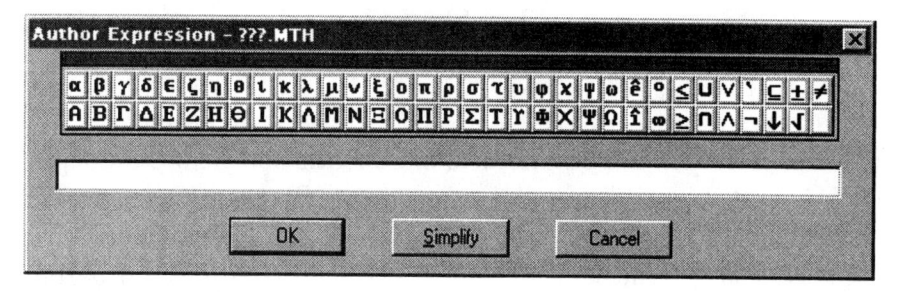

Try entering the expression
$$2+5=$$
and pressing [ENTER]. **DERIVE** will place the following output in the **Algebra** window.

```
#1:    2 + 5 = 7
```

Typing the "=" sign prompts **DERIVE** to perform the calculation. You might want to try this again without typing the "=" sign. That is, simply access the **Author Expression** window (as above), type
$$2+5$$
and press [ENTER].

```
#1:    2 + 5 = 7
#2:    2 + 5
```

DERIVE can do arithmetic on formulas entered on an input line. The standard arithmetic operations are $+, -, *, /,$ and $\char`\^$ for addition, subtraction, multiplication, division and exponentiation. However, multiplication is "implied", so the "$*$" sign is optional on input lines. For example, authoring the expressions
$$(2+4)(3-1)=$$
and
$$(2+4)*(3-1)=$$
leads to the same output in the **Algebra** window.

```
#3:    (2 + 4)·(3 - 1) = 12
#4:    (2 + 4)·(3 - 1) = 12
```

The standard order of operations is exponentiation before multiplication and division, and then addition and subtraction. To be safe, use parentheses to be sure that the operations are performed in the desired order. For example, (3+4)/7 is not the same as 3+4/7. Be sure to use round parentheses

1.1. DERIVE AS A CALCULATOR

rather than square brackets [ ] or curly braces { }, which have other meanings in **DERIVE**.

DERIVE knows a large number of standard mathematical functions including sine, cosine, tangent, inverse sine, inverse cosine, inverse tangent, secant, cosecant, cotangent, inverse secant, inverse cosecant, inverse cotangent, exponential, natural logarithm, square root and absolute value. The associated **DERIVE** commands are given in the table below.

SIN	**COS**	**TAN**	**ASIN**
ACOS	**ATAN**	**SEC**	**CSC**
COT	**ASEC**	**ACSC**	**ACOT**
EXP	**LN**	**SQRT**	**ABS**

DERIVE also knows the numbers π and e. The number π can be accessed by either entering [Ctrl]+p or entering **pi**. Similarly, the number e can be accessed by entering [Ctrl]+e. You should also be aware that the number e appears with a "hat" in the **Algebra** window.

It is important to understand the distinction between numbers that **DERIVE** treats as exact values, such as 2, 1/3, $\sqrt{2}$, e and π, and those that it treats as *floating-point decimal values* such as 2.0, .33333, 1.414, and 3.14. The number 1/3 is an expression that represents the exact value of one-third, whereas .33333 is a floating-point decimal approximation of 1/3. It is possible to force DERIVE to either treat every value as exact or to treat every value as approximate. This can be done by clicking on the **Declare** menu item, and then selecting **Algebra State**,

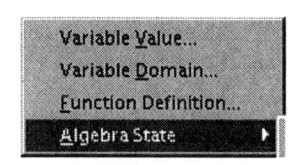

```
Variable Value...
Variable Domain...
Function Definition...
Algebra State          ▶
```

and finally **Simplification**.

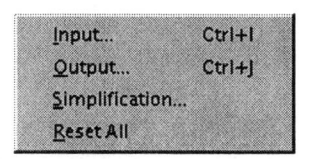

```
Input...          Ctrl+I
Output...         Ctrl+J
Simplification...
Reset All
```

It is then a simple matter to select the appropriate **Precision Mode**.

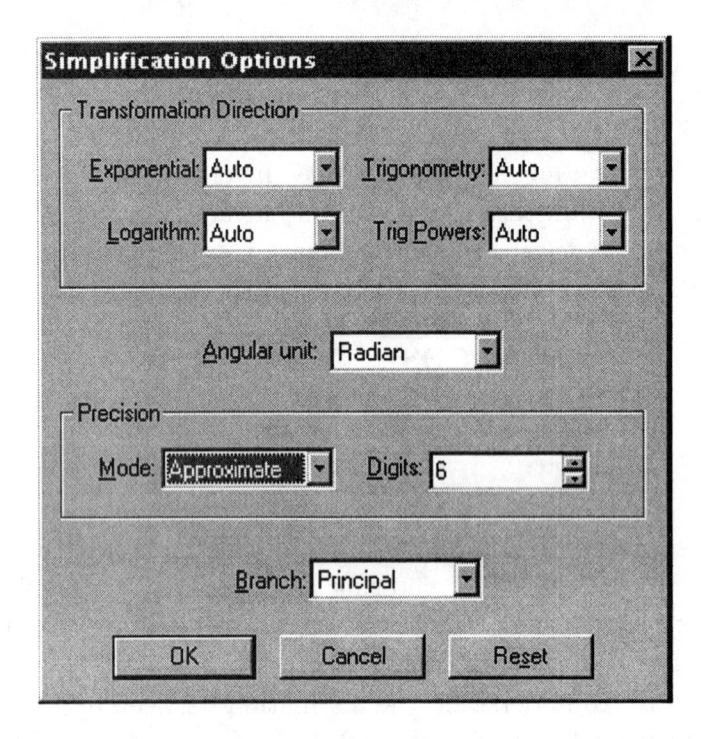

Press [OK] (twice) after making the selection. If you make a selection which is different from the current **Mode**, then you will notice a change in the **Algebra** window.

```
#4:   (2 + 4)·(3 - 1) = 12
#5:   Precision := Approximate
```

You will also notice a change in the type of output that you obtain. For example, if you **Author** the **Expression**

$$3/7 =$$

in **Approximate Mode**, then you will see

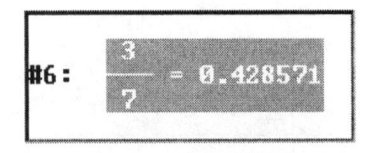

$$\#6: \quad \frac{3}{7} = 0.428571$$

Whereas, if the **Mode** is reset to **Exact** then you will see

1.2. ASSIGNING VARIABLES

```
#7:   Precision := Exact

#8:    3     3
      --- = ---
       7     7
```

Finally, you should know that it is possible to obtain approximate results while in **Exact Mode** by clicking on the [≈] icon in the tool bar of the **DERIVE** window or by pressing [Ctrl]+G [ENTER].

1.2 Assigning Variables

DERIVE answers are often used again in subsequent calculations and therefore **DERIVE** provides a way to store and recall earlier results. One way to refer to an earlier result is to refer to the line number of the output. For example, suppose the calculation of $21^6 + 1$ is done in two steps as follows. First author the expression

$$21\hat{}6$$

Then author the expression

$$\#1 + 1 =$$

The result is shown below.

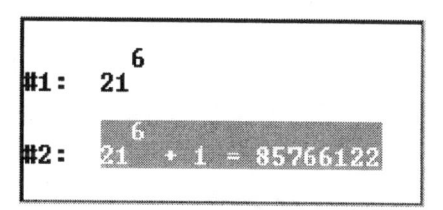

```
#1:   21^6

#2:   21^6 + 1 = 85766122
```

Notice that **DERIVE** interprets "#1" as the output in #1.

Now let's suppose we want to take the output on the right hand side of the equation of #2 and use it in a calculation. For example, if we want to take the reciprocal of this value then we can author the expression

$$1/\text{rhs}(\#2) =$$

and obtain the following result in the **Algebra** window.

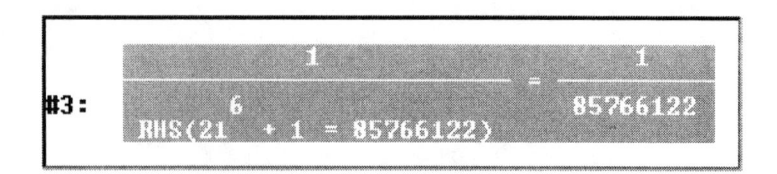

$$\#3: \quad \frac{1}{6} = \frac{1}{85766122}$$
$$\text{RHS}(21 \div 1 = 85766122)$$

If we author the same expression but click on the **Simplify** button to leave the **Author Expression** window, then we only see the output from our calculation.

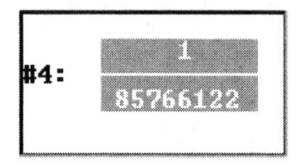

$$\#4: \quad \frac{1}{85766122}$$

Names or labels can also be used to store and refer to results. For example, the number 22/79+34/23 can be assigned to the variable *a* by authoring the expression

$$a := 22/79 + 34/23$$

and obtaining the output

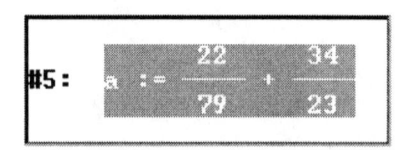

$$\#5: \quad a := \frac{22}{79} + \frac{34}{23}$$

If we want to see the value of **a** then we can author the expression

$$a =$$

and obtain

$$\#6: \quad a = \frac{3192}{1817}$$

(Note that we are clearly in **Exact Mode**. See Section 1).

We can also perform a variety of calculations using this variable. For example, to compute $(3192/1817)^2$, we simply author the expression

$$a^\wedge 2 =$$

and obtain

1.2. ASSIGNING VARIABLES

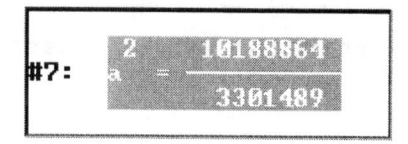

#7: $\quad a^2 = \dfrac{10188864}{3301489}$

To compute $1/a$ we author the expression

$$1/a =$$

and obtain

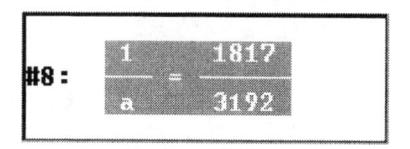

#8: $\quad \dfrac{1}{a} = \dfrac{1817}{3192}$

Finally, to compute $\sqrt{a}$, in **Approximate Mode**, we change simplification modes (see above) and author the expression

$$\text{SQRT(a)} =$$

#9: Precision := Approximate
#10: $\sqrt{a} = 1.32542$

To more easily distinguish between various labels, use descriptive names. For example, the retail price and wholesale cost of an item can be stored by authoring the expressions

$$\text{price:=4.95}$$

and

$$\text{cost:=2.80}$$

#11: price := 4.95
#12: cost := 2.8

The value of the profit is then given by authoring the expression

$$\text{profit} := \text{price} - \text{cost}$$

and clicking the ▄ icon (or pressing [Ctrl]+B [ENTER]).

```
#13: profit := price - cost
#14: 2.15
```

Note that labels are not case sensitive unless this option is selected. Changes can be made in this option by clicking on the **Declare** menu, and then selecting **Algebra State** and **Input**. The following window will appear.

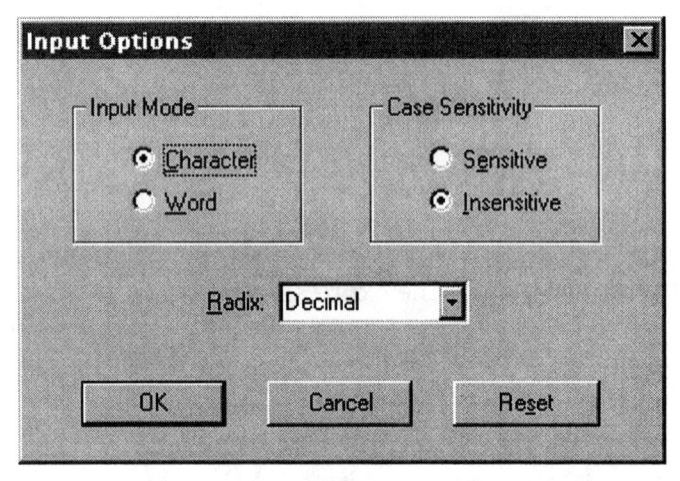

[Input Options dialog box with Input Mode (Character/Word), Case Sensitivity (Sensitive/Insensitive), Radix: Decimal, and OK/Cancel/Reset buttons]

It is much easier to author expressions if the "case" is left "insensitive". We will use this convention throughout the remainder of this manual.

A variable keeps its value until it is assigned a new value or until it is cleared (or unassigned). The value from the assigned variable **cost** can be cleared by entering authoring the expression

$$\text{cost} :=$$

Now the variable **cost** has no value assigned to it. You can check this by authoring the expression

$$\text{cost} =$$

```
#15: cost :=
#16: cost = cost
```

As a final example, enter an expression that describes the area of a circle of radius r.

$$\text{area} := \text{pi} * r\hat{\ }2$$

```
#17:  area := π·r²
```

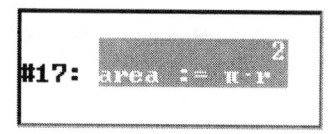

Note that π can also be entered by pressing [Ctrl]+P. To evaluate this area when $r = 5$, author the expression

$$r := 5$$

The value $r = 5$ will automatically be substituted into **area**, and we can see the value of area by authoring the expression

$$\text{area} =$$

```
#18:  r := 5
#19:  area = 78.5398
```

1.3 Algebra Commands

We have seen how to manipulate numbers and assign them to variables (or labels). **DERIVE** can also manipulate algebraic expressions involving labels or variables.

For example, to expand the expression $(3x - 2)^2(x^3 + 2x)$, author the expression

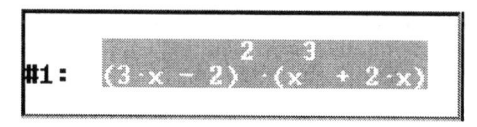

$$(3x - 2)\hat{\ }2(x\hat{\ }3 + 2x)$$

```
#1:  (3·x - 2)²·(x³ + 2·x)
```

This expression can be expanded a few different ways.. The first of these is to click on the **Simplify** menu item, and then select **Expand** and press [ENTER].

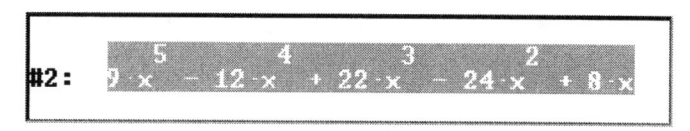

```
#2:  9·x⁵ - 12·x⁴ + 22·x³ - 24·x² + 8·x
```

The other way to expand this expression to to use **DERIVE**'s **EXPAND** command. Author the expression

$$\text{EXPAND}(\#1)$$

#3: EXPAND((3·x - 2)^2·(x^3 + 2·x))

and then either click on the ▭ icon or press [Ctrl]+B [ENTER].

#4: 9·x^5 - 12·x^4 + 22·x^3 - 24·x^2 + 8·x

Notice that this entire process can also be combined into one step. Simply author the expression

$$\text{EXPAND}((3x - 2)\hat{\ }2(x\hat{\ }3 + 2x))$$

and then click on the ▭ icon or press [Ctrl]+B [ENTER]. It is also possible to simply highlight the expression in the **Algebra** window and press [Ctrl]+E [ENTER].

DERIVE can factor the polynomial $x^6 - 1$ in a number of different ways. Since these are all similar to the methods above for expanding and expression, we only demonstrate one way of doing this. Simply author the expression

$$\text{FACTOR}(x\hat{\ }6 - 1)$$

and press [Ctrl]+B [ENTER].

#5: FACTOR(x^6 - 1)

#6: (x + 1)·(x - 1)·(x^2 + x + 1)·(x^2 - x + 1)

DERIVE can also simplify algebraic expression. For example, to simplify the expression

$$\frac{x^2 - x}{x^3 - x} - \frac{x^2 - 1}{x^2 + x}$$

1.4. PLOTS

author the expression

$$(x^2 - x)/(x^3 - x) - (x^2 - 1)/(x^2 + x)$$

followed by [Ctrl]+B [ENTER].

$$\#7: \quad \frac{x^2 - x}{x^3 - x} - \frac{x^2 - 1}{x^2 + x}$$

$$\#8: \quad -\frac{x^2 - x - 1}{x \cdot (x + 1)}$$

As above, this process can also be performed using pull down menus.

Note: **DERIVE** has an on-line help facility that is invoked by clicking on the **Help** menu item. Help with a specific command can be obtained by selecting the **Index** under **Help**.

1.4 Plots

Plots are extremely easy to create in **DERIVE**. Let's consider an example. Suppose we want to plot the graph of

$$y = \frac{2x^2 - 4}{x + 1}$$

over the interval $-6 \leq x \leq 6$. Start by authoring the expression

$$y = (2x^2 - 4)/(x + 1)$$

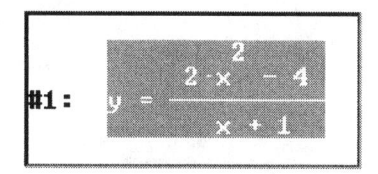

$$\#1: \quad y = \frac{2 \cdot x^2 - 4}{x + 1}$$

Then highlight this expression (it will still be highlighted if you have not authored any further expressions) and press on the **2-D Plot** icon. This will create a **2-D Plot** window similar to the one below.

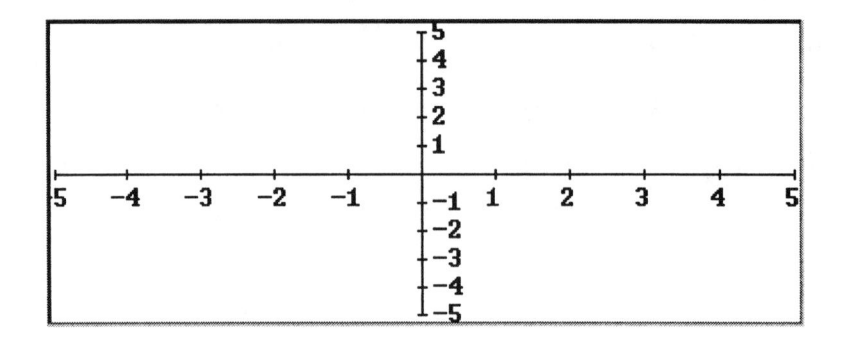

Do not be concerned that there is no plot showing in this window. Simply click on the **Plot!** menu item and obtain

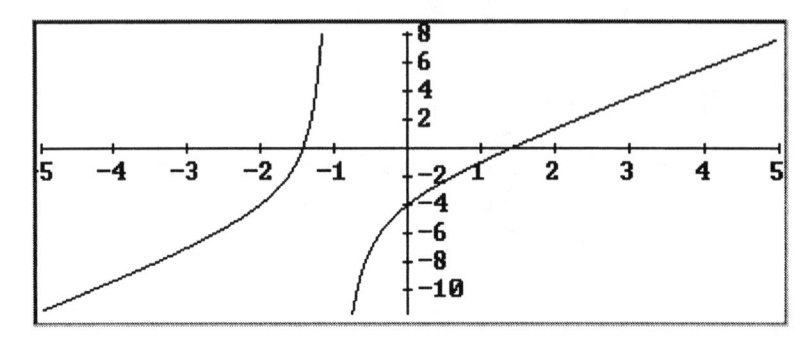

Now, recall that we originally requested a plot on the interval $-6 \leq x \leq 6$. To obtain this plot simply click on the **Set** menu item (make sure the plot window is active) and select **Range**. The window below is the result of these actions and changes made to the **Left**, **Right**, **Bottom** and **Top** input boxes (corresponding to xmin, xmax, ymin and ymax respectively).

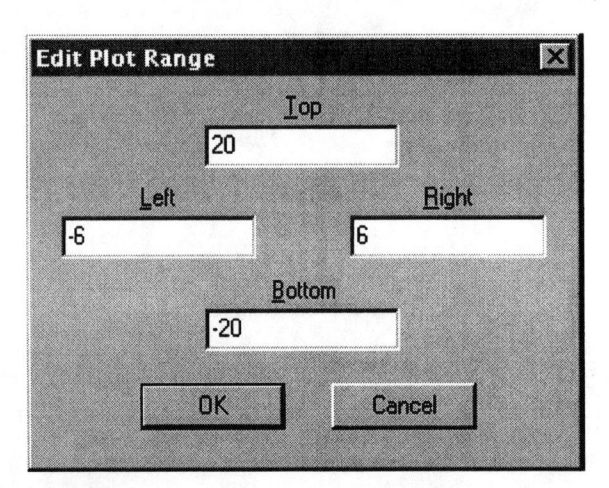

1.4. PLOTS

Click [OK] or press [ENTER] and the **2-D Plot** window will immediately be updated.

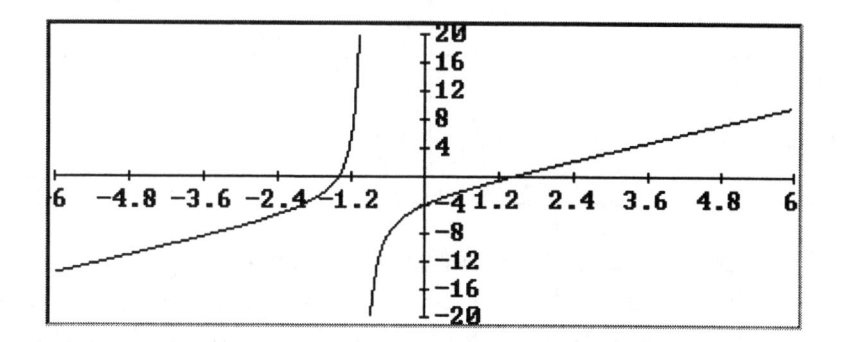

By changing the plot range, different aspects of the graph can be viewed. For example, changing the x- and y-ranges to $x = -50..50$, $y = -100..100$ displays the graph for larger values of x, and we can see that the graph of the function approaches the line $y = 2x-2$ as a skewed asymptote. However, with such large values of x, the vertical asymptote at $x = -1$ becomes obscured.

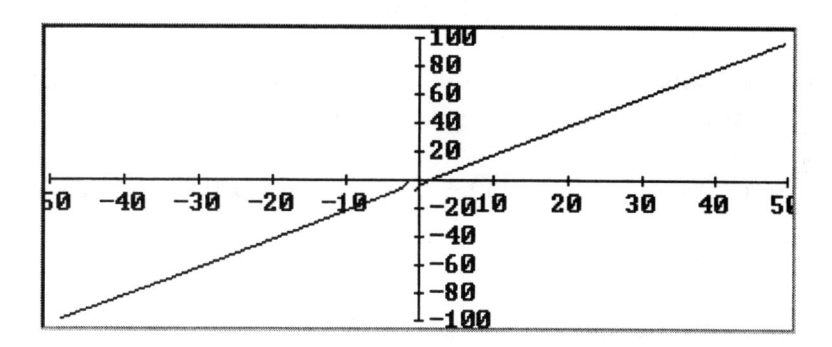

More than one expression can be graphed on the same plot by creating the plots one at a time. For example, reset the **Range** settings above to give the plot of $y = \frac{2x^2-4}{x+1}$ for $-6 \leq x \leq 6$ and $-20 \leq y \leq 20$. Now, return to the **Algebra** window without closing the **2-D Plot** window (just click on the **2-D Plot** window). If we want a plot of $y = 2x - 2$ to appear in the plot window we have already created, then we simply author the expression

$$y = 2x - 2$$

click on the **2-D Plot** icon and click on the **Plot!** menu item.

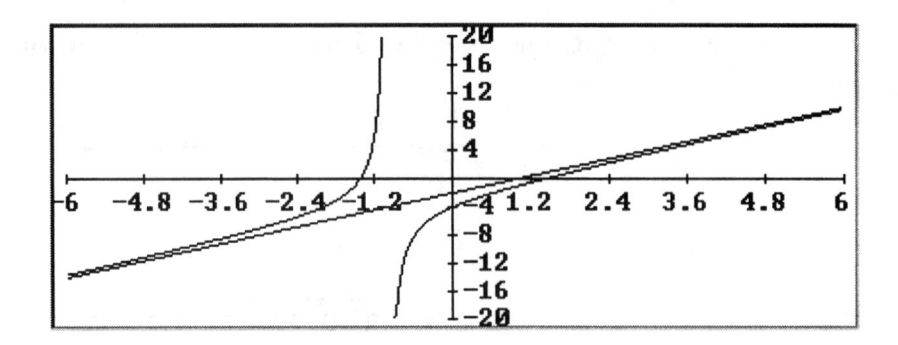

This method can be repeated several times to place a large number of plots in a window. Note that if the expressions have already been authored and appear earlier in the **Algebra** window, simply locate, highlight and plot each one separately.

Remark 1 *It is only necessary to author the expressions which appear on the right hand side of the equations in each case above. For example, to plot* $y = 2x - 2$, *it is only necessary to author the expression*

$$2x - 2$$

Another method for plotting *three or more* expressions is to place them in a vector. For example, if we want to plot each of

$$\sin(x), \ \cos(2x), \ 2\sin(x) + \cos(3x) \text{ and } \cos(x) - \sin(x)$$

on the same graph then we can author the expression

```
[SIN(x),COS(2x),2SIN(x)+COS(3x),COS(x)-SIN(x)]
```

highlight it and create a plot in the manner shown above. The output is shown below.

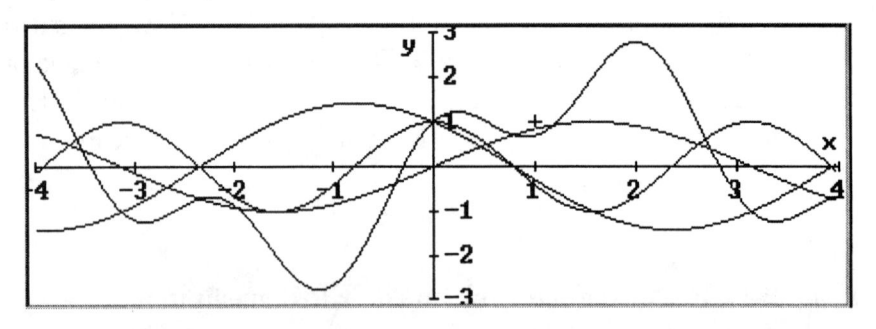

1.4. PLOTS

This plot can also be obtained by authoring the expressions

$$SIN(x)$$

$$COS(2x)$$

$$2SIN(x)+COS(3x)$$

$$COS(x)-SIN(x)$$

separately, and then referring to them in a vector using their output numbers. For example, if these expressions are authored as shown below,

```
#4:   SIN(x)
#5:   COS(2·x)
#6:   2·SIN(x) + COS(3·x)
#7:   COS(x) - SIN(x)
```

then we can author the vector needed to create the plot by authoring the expression

$$[\#4, \#5, \#6, \#7]$$

Then we can simply highlight, click the **2-D Plot** icon and click on the **Plot!** menu item.

There is a simple reason why this process will not create the plot of 2 functions. Namely, **DERIVE** will treat an expression of the form $[\cos(x), \sin(x)]$ as a parametric plot. However, there is a simple trick for using vectors to create plots of 2 expressions. For example, if you want to plot the expressions $y = x^2 - 1$ and $y = 2x$ on the same graph then you can author the expression

$$[0, \texttt{x\^{}2-1}, 2x]$$

and create a plot as above. Then **DERIVE** treats this as a request to plot 3 expressions. In this case, the first expression is just the x-axis.

Finally, plots can be deleted by using the **Edit** menu (while the **2-D Plot** window is active) or by closing the **2-D Plot** window. Additional plot topics will be discussed throughout this manual.

Exercises:

1. Assign the variable name a to the number $2\pi/5$ and then set **DERIVE**'s **Precision Mode** to compute the decimal approximations for a^2, $1/a$, $\sqrt{a}$, $a^{1.3}$, $\sin(a)$, and $\tan(a)$ to 8 decimal places.

2. Repeat Exercise 1 with 20 significant digits by using the **APPROX** command. For example, to calculate a^2 to 20 significant digits, author the expression
$$\texttt{APPROX(a\^{}2,20)}$$
and then press [CTRL]+B [ENTER].

3. Expand the following expressions.

    (a) $(x^2 + 2x - 1)^3 (x^2 - 2)$

    (b) $(x + a)^5$

    Note: If you have done Exercise 1 or 2, the label a already has a value assigned to it. Recall that this value should be unassigned by authoring the expression `a:= `.

4. Factor the expression $x^2 + 3x + 2$.

5. Try factoring $x^2 + 3x - 11$. What is the problem?

6. Factor $x^8 - 1$.

7. Simplify
$$\frac{2x^2}{x^3 - 1} + \frac{3x}{x^2 - 1}$$

8. Plot the graph of $y = \tan(x)$. Experiment with the x- and y-ranges to obtain a reasonable plot of one period of $y = \tan(x)$.

9. (a) Plot the graph of $y = \dfrac{3x^2 - 2x + 1}{x - 1}$ over a small interval containing $x = 1$, for example, $0 \leq x \leq 2$. Experiment with the y-range to obtain a reasonable plot. What happens to the graph near $x = 1$?

1.4. PLOTS

(b) Now plot the same expression over a large interval such as $-100 \leq x \leq 100$. Note that the behavior of the graph near $x = 1$ is no longer apparent. Why do you think this happens?

10. Plot the expressions $\sin(x)$, $\sin(2x)$, and $\sin(4x)$ over the interval $0 \leq x \leq 2\pi$ on the same coordinate axes. Now plot the same expressions over the interval $0 \leq x \leq 4\pi$. Delete your plots (but keep the **2-D Plot** window active) and click on the **Options** menu, and select **Autoscale Mode**. Finally, click on the **Plot!** menu. What happened?

11. Compute 21! using **DERIVE**.

12. Give the first 40 digits of π. (see exercise 2)

13. Compute the exact and floating-point values of $\sin(\pi/4)$.

14. Compute the exact and floating-point values of $\sin(1)$.

15. Compute the number of seconds in one year, showing the units in your product as each factor is entered.

16. Just as the command **FACTOR** will decompose a polynomial into irreducible factors, it will also give the prime decomposition of an integer. Use the **FACTOR** command to show that $2^{23} - 1$ is not prime. Note: Make sure the **Precision Mode** is set to **Exact**.

17. See what happens when the command **EXPAND** is applied to $(a+b)/c$.

18. Use **EXPAND** to expand the expression $\ln\left(\frac{ab}{c}\right)$. If **DERIVE** responds with the same expression, then click on the **Declare** menu, select **Algebra State** and **Simplification Options**. Then change the setting for logarithms to **Expand**.

19. Factor the expression $e^{2x} - 1$, by first using the **FACTOR** command. Remember to use EXP(2x) for e^{2x}.

20. Author the expressions P:=[2,1,4,6] and Q:=[3,7,-2,1]. Then, author the following expressions:

$$p \text{ sub } 1 =$$

$$p \text{ sub } 3 =$$

$$q \text{ sub } 4 =$$

What does the **SUB** command do?

21. Plot the expressions $-0.939x - 5.14$ and $-3.14x + 17.32$ on the same graph. Set the plot range so that the intersection point appears in the plot. Click on the intersection point with the mouse and read the approximate coordinates of the intersection point in the lower left hand corner of the window. Record these values.

22. Set the **Precision Mode** to **Approximate** with 6 decimal places of accuracy. Author the expression

    SOLVE([y=-x/2+5/2,y=-3x+5],[x,y])

    and press [Ctrl]+B [ENTER]. Compare your result with problem 21.

23. Repeat problem 22 after setting the **Precision Mode** to **Exact**.

Chapter 2

Vectors

This chapter introduces a number of capabilities of **DERIVE** for visualizing and working with vectors.

- Related material is found in Chapter 9 of Stewart's **Multivariable Calculus: Concepts and Contexts**.

2.1 Vector Basics

There are two ways to create vectors in **DERIVE**. One way to create a vector is to author a vector and make use of the dialog boxes which appear.

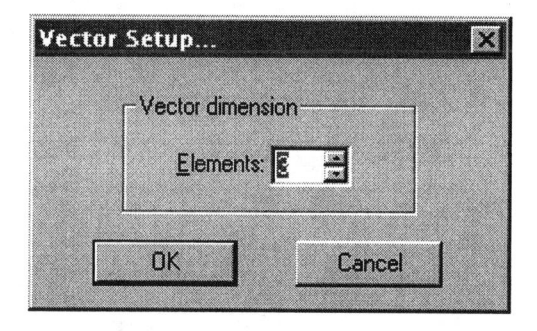

First, input the number of entries in the vector and press **OK**. The screen below will appear. Enter the values shown and press **OK**.

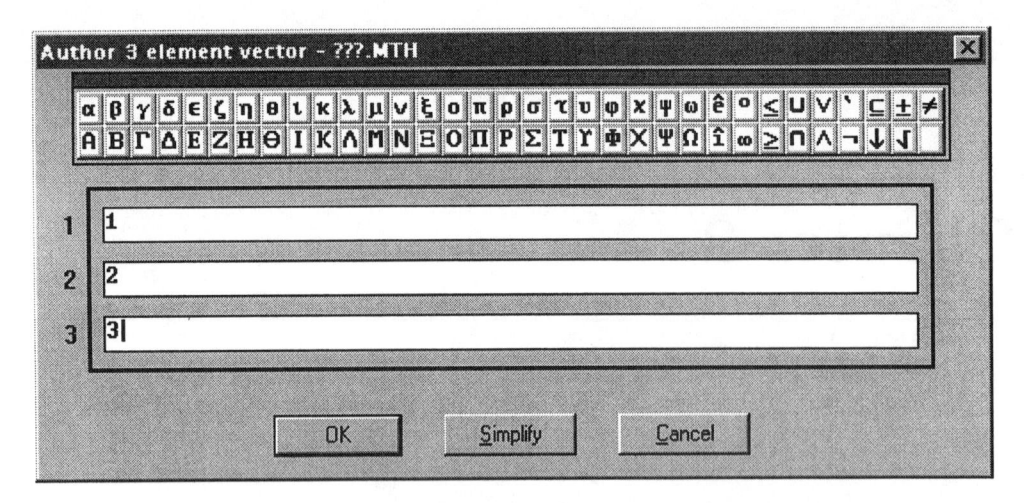

[Dialog box: Author 3 element vector - ???.MTH, with entries 1, 2, 3 and OK / Simplify / Cancel buttons]

The other way to create a vector in **DERIVE** is to author an expression and enter the values in the vector enclosed by (square) brackets and separated by commas.

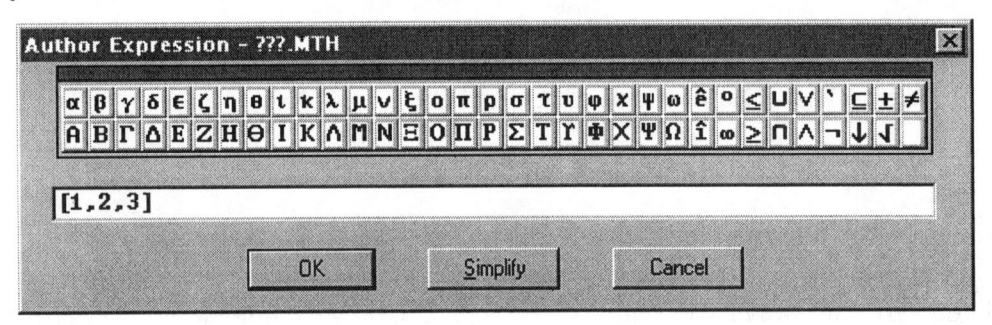

[Dialog box: Author Expression - ???.MTH, with entry [1,2,3] and OK / Simplify / Cancel buttons]

The **Algebra Window** below shows that the results of these methods are identical.

```
#1:   [1, 2, 3]
#2:   [1, 2, 3]
```

Note that **DERIVE** makes no distinction between row vectors and column vectors.

In our discussions we will follow the convention in Stewart's text of using "angle brackets" to enclose vector coordinates (e.g., $\langle 1, 2, 3 \rangle$) in order to distinguish vectors from points.

Vectors are really just special *matrices* having either only one row or only one column. Like vectors, matrices can be created in two ways. One way to

2.1. VECTOR BASICS

create a matrix is to author a matrix and follow the dialog boxes as above. The other way to create a matrix is to author an expression as shown below.

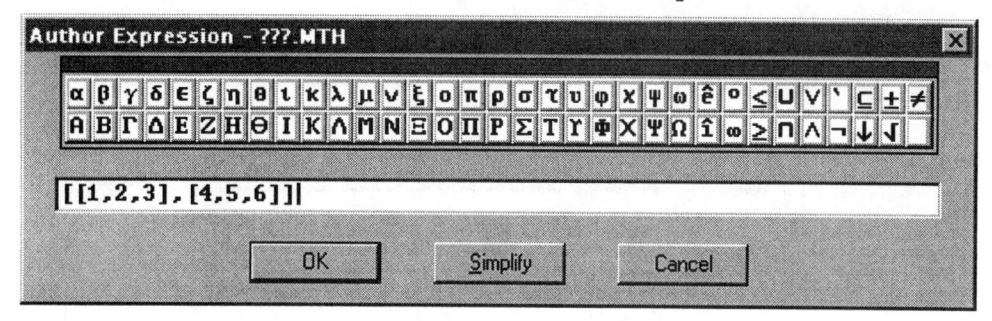

Clicking **OK** results in the following entry in the **Algebra Window**.

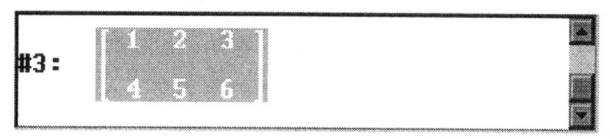

#3: $\begin{bmatrix} 1 & 2 & 3 \\ 4 & 5 & 6 \end{bmatrix}$

Vectors can be plotted in **DERIVE** by simply plotting the line segment which represents the vector. For example, suppose we want to plot the vector $\langle 2, 2 \rangle$. One way to visuallize this vector is to plot the line segment connecting the points $(0,0)$ and $(2,2)$. We start by creating a matrix which contains this information. Author the expression

$$[[0, 0], [2, 2]]$$

to obtain

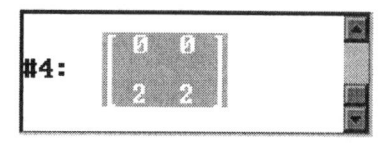

#4: $\begin{bmatrix} 0 & 0 \\ 2 & 2 \end{bmatrix}$

Now click on the **2D-Plot** icon to obtain the plot below.

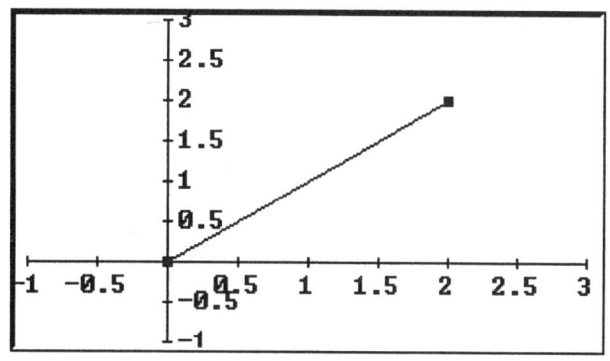

In general, if we have a point P and a vector **u**, and we want to visualize the vector **u** with its initial end at the point P then we can plot the line segment from P to $P+u$ (where u denotes the point for which **u** is a position vector). For example, suppose we want to visualize the vector $\mathbf{u} = \langle 2, 2 \rangle$ with initial end at $P = (1, -2)$. Start by deleting the plot above. Then author the expressions #5, #6 and #7 shown below, followed by pressing [=].

```
#5:   p := [-1, 2]
#6:   v := [2, 2]
#7:   [p, p + v]
#8:   [ -1  2 ]
      [  1  4 ]
```

Finally, click on the **2D-Plot** icon and then click on **Plot!** to obtain the sketch below.

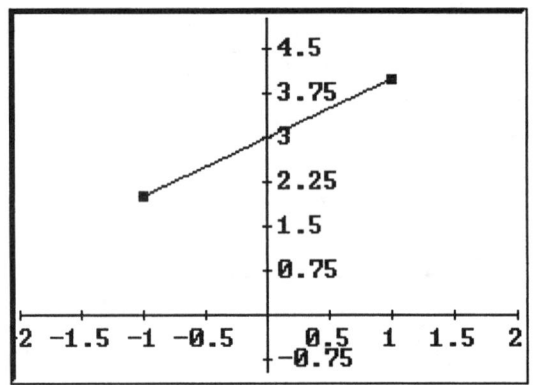

We conclude this section by demonstrating how **DERIVE** can be used to refer to entries in a vector or matrix. Consider the vector $\langle 2, 3, -1 \rangle$ shown in the **Algebra** window below. The **SUB** command can be used to refer to entries in this vector by authoring the expression

#1 SUB 2 =

```
#1:   [2, 3, -1]
#2:   [2, 3, -1]  = 3
              2
#3:   [2, 3, -1]  = 3
              2
```

Entries can also be extracted from vectors which have been assigned variable names, and from matrices by repeated use of the **SUB** command.

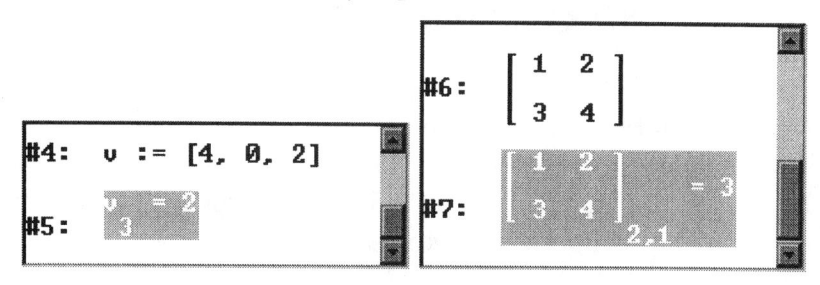

```
#4:   v := [4, 0, 2]

#5:   v = 2
      3
```

```
#6:   [ 1  2 ]
      [ 3  4 ]

#7:   [ 1  2 ]    = 3
      [ 3  4 ]
              2,1
```

We Authored the Expressions below to create #5 and #7 in the **Algebra** window shown above.

$$v \text{ SUB } 3 =$$

$$\#6 \text{ SUB } 2 \text{ SUB } 1 =$$

2.2 Length and Arithmetic Operations

- Detailed discussion and definitions of the notions in this section are found in Section 9.2 of Stewart's **Multivariable Calculus: Concepts and Contexts.**

The length of a two-dimensional vector $\mathbf{a} = \langle a_1, a_2 \rangle$ is

$$|\mathbf{a}| = \sqrt{a_1^2 + a_2^2}.$$

Similarly, the length of a three-dimensional vector $\mathbf{a} = \langle a_1, a_2, a_3 \rangle$ is

$$|\mathbf{a}| = \sqrt{a_1^2 + a_2^2 + a_3^2}.$$

DERIVE has a built-in function for computing the length of a vector. This function is **abs**. The **abs** function can be applied to either numeric or symbolic expressions. For example, the norm of the vector $\langle 2, -1 \rangle$ can be obtained by authoring the expression

$$\mathbf{abs}([2, -1]) =$$

Similarly, the norm of the vector $\langle x, y, z \rangle$ can be computed by authoring the expression

$$\mathbf{abs}([x, y, z]) =$$

The results are shown in the **Algebra** window below.

```
#1:  |[2, -1]| = √5
#2:  |[x, y, z]| = √(x² + y² + z²)
```

Addition and subtraction of vectors and multiplication of a vector by a scalar are all handled automatically by **DERIVE**. For example, suppose we author the expressions in #3, #4 and #5 shown below. Then **u − 2v + 3w** can be computed by authoring the expression

$$u - 2*v + 3*w =$$

```
#3:  u := [3, -1, 2]
#4:  w := [1, 2, -1]
#5:  v := [1, 1, -3]
#6:  u - 2·v + 3·w = [4, 3, 5]
```

2.3 Dot Products and Projections

- Detailed discussion and definitions of the notions in this section are found in Section 9.3 of Stewart's **Multivariable Calculus: Concepts and Contexts**.

Dot products can be computed in **DERIVE** by either using the standard multiplication operator or the period. We demonstrate this below. We start by authoring the expressions in #1 and #2 below.

```
#1:  u := [-1, 1, 3]
#2:  v := [2, 2, -1/2]
```

The the dot product of **u** and **v** can be computed by authoring either of the following expressions:

$$u * v =$$
$$u . v =$$

The **Algebra** window below verifies that the calculations are the same.

2.3. DOT PRODUCTS AND PROJECTIONS

```
#3:   u·v = - 3/2

#4:   u·v = - 3/2
```

The cosine of any angle θ formed by two position vectors **a** and **b** is

$$\cos\theta = \frac{\mathbf{a}\cdot\mathbf{b}}{|\mathbf{a}|\,|\mathbf{b}|}\,.$$

If we take θ to be the smaller of the two angles formed by **a** and **b**, i.e., the angle *between* **a** and **b**, then

$$\theta = \cos^{-1}\left(\frac{\mathbf{a}\cdot\mathbf{b}}{|\mathbf{a}|\,|\mathbf{b}|}\right).$$

Let's define a **DERIVE** function named **ANGLE** that can be used to compute θ. The function **ANGLE** is defined in #6 below. #7 is the result of applying **ANGLE** to the vectors **u** and **v** defined above. The value in #8 (given in radians) is obtained by clicking [≈] after authoring the expression in #7.

```
#6:   ANGLE(a, b) := ACOS( a·b / (|a|·|b|) )

#7:   ANGLE(u, v)

#8:   1.72891
```

One of the most important uses of the dot product is the calculation of the projection of one vector onto another. Let **a** and **b** be two vectors (two- or three-dimensional). The (vector) projection of **a** onto **b** is the vector that has the same direction as **b** and together with **a** forms a right triangle. The **component of a along b** is

$$\mathrm{comp}_\mathbf{b}\mathbf{a} = \frac{\mathbf{b}\cdot\mathbf{a}}{|\mathbf{b}|}\,,$$

and the **projection of a onto b** is

$$\text{proj}_b a = (\text{comp}_b a) \frac{b}{|b|}.$$

Be careful to note that the component of **a** along **b** is a scalar, while the projection of **a** onto **b** is a vector. The component of **a** along **b** is in fact either the length of the projection of **a** onto **b** or the negative of that length.

We can create functions **COMP** and **PROJ** in **DERIVE** by authoring the expressions below.

$$\text{COMP}(a, b) := a * b / \text{abs}(b)$$

$$\text{PROJ}(a, b) := a * b / \text{abs}(b)\verb|^|2 * b$$

The following **Algebra** window shows the result of their definition.

```
#9:   PROJ(a, b) := (a·b)/|b|² · b

#10:  COMP(a, b) := (a·b)/|b|
```

Example 1 *For the vectors* $a = \langle 2, 2, -1 \rangle$ *and* $b = \langle 5, 1, -3 \rangle$, *find the component of* **a** *along* **b** *and the projection of* **a** *onto* **b**. *Then find the component of* **b** *along* **a** *and the projection of* **b** *onto* **a**.

We being by defining the vectors **a** and **b**.

```
#11:  a := [2, 2, -1]
#12:  b := [5, 1, -3]
```

Then the component of **a** onto **b** and the projection of **a** onto **b** can be computed as shown in the **Algebra** window below.

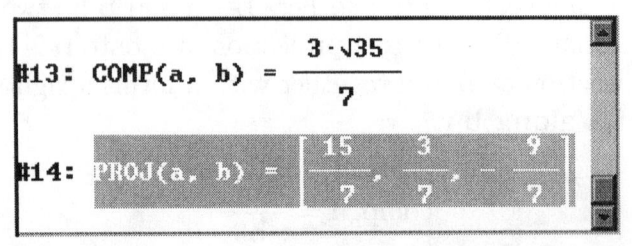

```
#13:  COMP(a, b) = (3·√35)/7

#14:  PROJ(a, b) = [15/7, 3/7, -9/7]
```

2.3. DOT PRODUCTS AND PROJECTIONS

the component of **b** onto **a** and the projection of **b** onto **a** can be computed in a similar manner.

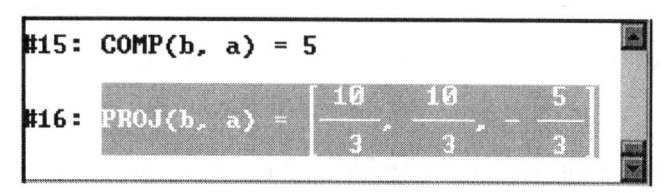

```
#15: COMP(b, a) = 5

#16: PROJ(b, a) = [ 10/3 , 10/3 , -5/3 ]
```

Exercises:

1. Create a matrix containing row vectors of the form $\langle \cos(k\pi/6), \sin(k\pi/6) \rangle$ for $k = 1, 2, \ldots, 12$. Then plot these vectors in the same **2-D Plot** window.

2. Edit the matrix in #1 so that the rows are of the form
$$\langle 2\cos(k\pi/6), \sin(k\pi/6) \rangle.$$
Then plot the vectors in the same **2-D Plot** window.

3. Edit the matrix in #2 so that the rows are of the form
$$\left\langle \frac{2}{k}\cos(k\pi/6), \frac{1}{k}\sin(k\pi/6) \right\rangle.$$
Then plot the vectors in the same **2-D Plot** window.

4. Let $\mathbf{a} = \langle 1, 2, 5 \rangle$ and $\mathbf{b} = \langle 1, 02, -3 \rangle$. Compute $|\mathbf{a}|$, $|\mathbf{b}|$, $\mathbf{b} \cdot \mathbf{b}$, the angle between **a** and **b**, the projection of **b** onto **a**, and the projection of **a** onto **b**.

5. Verify that the vectors $\langle t, \sin(t), 1 - t \rangle$ and
$$\langle t - 1, (t - 1)\cos(t), t - \cos(t)\sin(t) \rangle$$
are orthogonal for all real numbers t.

6. Verify that $\langle -s + t, s + 2t, t - 3s \rangle$ is orthogonal to $\langle 7, -2, -3 \rangle$ for all real numbers s and t.

7. Find each of the two-dimensional unit vectors that form an angle of $\pi/6$ radians with $\langle 2, 3 \rangle$.

2.4 The Cross Product

- Detailed discussion of the cross product of two vectors is found in Section 9.4 of Stewart's **Multivariable Calculus: Concepts and Contexts**.

The **CROSS** function in **DERIVE** computes the cross product of two three dimensional vectors.

```
#1:   u := [2, 2, -1]
#2:   v := [2, 0, 3]
#3:   CROSS(u, v) = [6, -8, -4]
```

The **CROSS** function also works on pairs of two-dimensional vectors. However, in this case, since the resulting vector is of the form $\langle 0, 0, z \rangle$, **CROSS** only returns the third component.

```
#4:   a := [1, -2]
#5:   b := [3, 1]
#6:   CROSS(a, b) = 7
```

Example 2 *Find a unit vector which is perpendicular to each of the vectors* $\mathbf{a} = \langle 1, 2, 3 \rangle$ *and* $\mathbf{b} = \langle -2, 1, 3 \rangle$.

The desired vector is found by computing the cross product of $\mathbf{a}$ and $\mathbf{b}$ and then dividing by its length. Clicking the [=] icon gives the result.

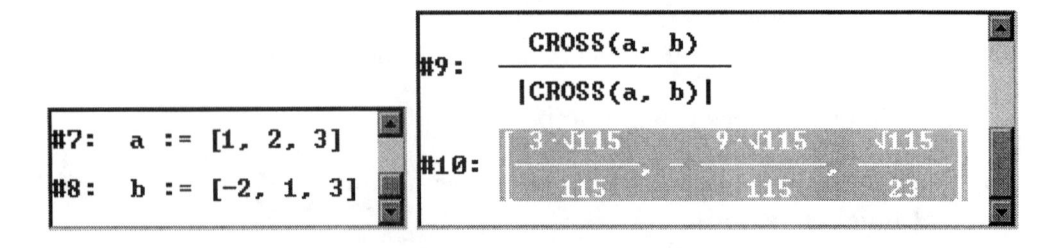

```
#7:   a := [1, 2, 3]
#8:   b := [-2, 1, 3]

#9:   CROSS(a, b)
      ─────────────
      |CROSS(a, b)|

#10:  [3·√115/115,  -9·√115/115,  √115/23]
```

Example 3 *Find a unit vector that is perpendicular to the plane that passes through the points* $(2, 3, 2)$, $(1, -1, 1)$, *and* $(4, -1, 0)$.

2.5. EQUATIONS OF LINES

The first step is to compute two vectors that are parallel to the plane. This can be done by forming displacement vectors using the three points above. Then we compute the cross product and divide by its length as in the previous example.

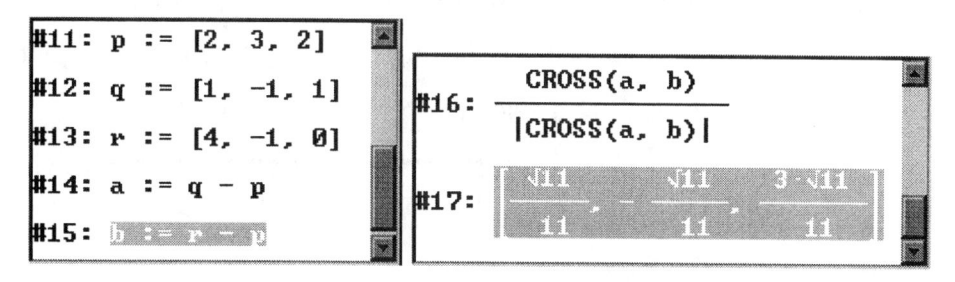

```
#11:  p := [2, 3, 2]
#12:  q := [1, -1, 1]
#13:  r := [4, -1, 0]
#14:  a := q - p
#15:  b := r - p
```

$$\#16: \quad \frac{\text{CROSS}(a, b)}{|\text{CROSS}(a, b)|}$$

$$\#17: \quad \left[\frac{\sqrt{11}}{11}, \frac{\sqrt{11}}{11}, \frac{3\sqrt{11}}{11}\right]$$

Recall that the length of the cross product of two vectors **a** and **b** is the area of the parallelogram spanned by **a** and **b**.

Example 4 *Find the area of the triangle whose vertices are the points*

$$A(3, 1, -1), B(7, 5, 1) \text{ and } C(3, 3, 3).$$

The area of the triangle is half that of the parallelogram determined by the position vectors of the points $B - A$ and $C - A$, and therefore half the length of the cross product of those vectors.

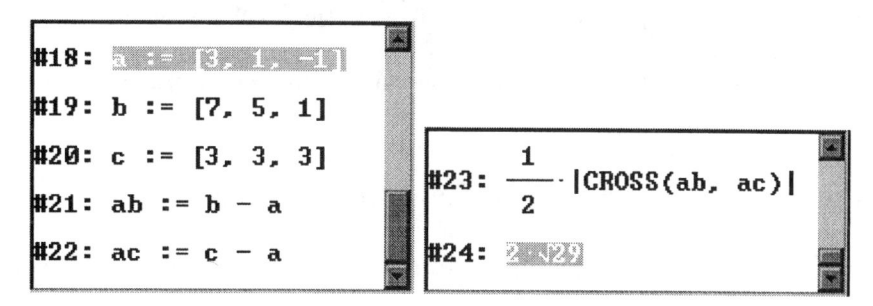

```
#18:  a := [3, 1, -1]
#19:  b := [7, 5, 1]
#20:  c := [3, 3, 3]
#21:  ab := b - a
#22:  ac := c - a
```

$$\#23: \quad \frac{1}{2} \cdot |\text{CROSS}(ab, ac)|$$

$$\#24: \quad 2\sqrt{29}$$

2.5 Equations of Lines

- A detailed discussion of the ideas in this section are found in Section 9.5 of Stewart's **Multivariable Calculus: Concepts and Contexts**.

Example 5 *Give a parameterization of the line that passes through the point $(1, 2)$ and has direction vector $\langle 3, 2 \rangle$. Then plot the line.*

A parameterization of the line is given by

$$\mathbf{r}(t) = \langle 1, 2 \rangle + t \langle 3, 2 \rangle.$$

We form the line as shown in the **Algebra** window on the left below. The last line in this window is formed by clicking on the ▢ icon. The plot is obtained by clicking on ▢ and selecting the **Plot!** menu item.

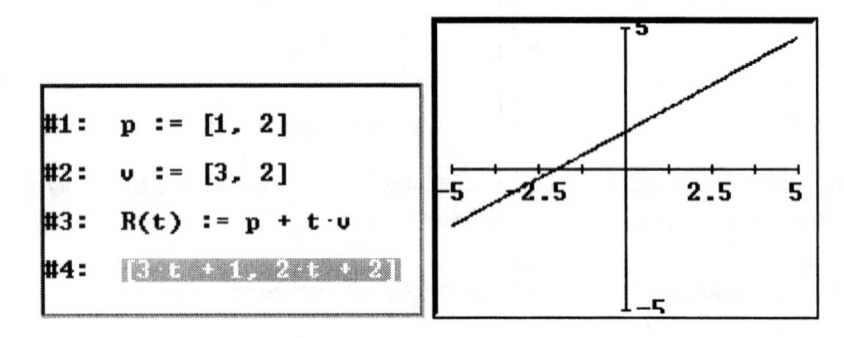

```
#1:  p := [1, 2]
#2:  v := [3, 2]
#3:  R(t) := p + t·v
#4:  [3·t + 1, 2·t + 2]
```

Intersecting Lines. Let's now consider the problem of determining the intersection of the two lines given by

$$\mathbf{r}(t) = \mathbf{r}_0 + t\,\mathbf{v} \text{ and } \mathbf{q}(t) = \mathbf{q}_0 + t\,\mathbf{w}.$$

Note that in three-dimensional space, two such lines need not intersect, even if they are not parallel.

Example 6 *Determine the intersection of the two lines given by*

$$\mathbf{r}(t) = \langle 1, 1, 0 \rangle + t \langle 2, -1/2, 1/3 \rangle \text{ and } \mathbf{q}(t) = \langle 0, 2, 0 \rangle + t \langle 1, -1/2, 1/9 \rangle.$$

The important thing to realize here is that the two lines may have a common point that corresponds to different parameter values. So the problem becomes that of finding values of t and s (if possible) such that

$$\mathbf{r}(t) = \mathbf{q}(s).$$

Our approach is to use the **SOLVE** command followed by clicking on the ▢ icon.

2.5. EQUATIONS OF LINES

```
#5:  R(t) := [1 + 2·t, 1 - t/2, t/3]

#6:  Q(s) := [s, 2 - s/2, s/9]

#7:  SOLVE([R(t) = Q(s)], [t, s])

#8:  [t = 1   s = 3]
```

Since the **SOLVE** command returned values for t and s, the lines must intersect. The point of intersection is shown below, along with verification of the intersection.

```
#9:  R(1) = [3, 1/2, 1/3]

#10: Q(3) = [3, 1/2, 1/3]
```

Example 7 *Show that the lines given by*

$$\mathbf{r}(t) = \langle 1 + 2t, t/3 - 5, t - 2 \rangle \text{ and } \mathbf{q}(t) = \langle 3t - 7, 1 - t/2, 5 - t \rangle$$

do not intersect.

Our goal is to show that the system of equations contained in

$$\mathbf{r}(t) = \mathbf{q}(s)$$

has no solution. We handle this example in the same manner as above. First we define the lines as $\mathbf{R}(t)$ and $\mathbf{Q}(t)$.

```
#11: R(t) := [1 + 2·t, t/3 - 5, t - 2]

#12: Q(t) := [3·t - 7, 1 - t/2, 5 - t]
```

Then we attempt to solve the equations $\mathbf{R}(t) = \mathbf{Q}(s)$ for t and s. Using **SOLVE** and clicking on the ▪ icon gives the **Algebra** window below. Notice the empty brackets in #14 indicating that there is no solution.

```
#13: SOLVE( [R(t) = Q(s)], [t, s])
#14: []
```

Consequently, the lines do not intersect.

2.6 Equations of Planes

- Detailed discussion of the ideas in this section are found in Section 9.5 of Stewart's **Multivariable Calculus: Concepts and Contexts.**

A plane in three dimensional space is determined by a point P_0 on it and a normal vector $\mathbf{n}$. Let $\mathbf{r}_0$ be the position vector of P_0 and let $\mathbf{r}$ denote the position vector of an arbitrary point (x, y, z) on the plane. Then

$$(\mathbf{r} - \mathbf{r}_0) \cdot \mathbf{n} = 0$$

This is the vector equation of the plane which is equivalent to each of

$$\mathbf{r} \cdot \mathbf{n} = \mathbf{r}_0 \cdot \mathbf{n} \text{ and } ax + by + cz = d,$$

where $\mathbf{n} = \langle a, b, c \rangle$ and $d = \mathbf{r}_0 \cdot \mathbf{n}$.

Example 8 *Find the equation of the plane that contains the points*

$$A\,(3, 2, -1)\,, B\,(0, 3, 1)\ and\ C\,(1, 1, 1)\,.$$

Let $\mathbf{a}$, $\mathbf{b}$ and $\mathbf{c}$ be the position vectors A, B and C respectively. Then two vectors parallel to the plane are $\mathbf{u} = \mathbf{a} - \mathbf{c}$ and $\mathbf{v} = \mathbf{b} - \mathbf{c}$.

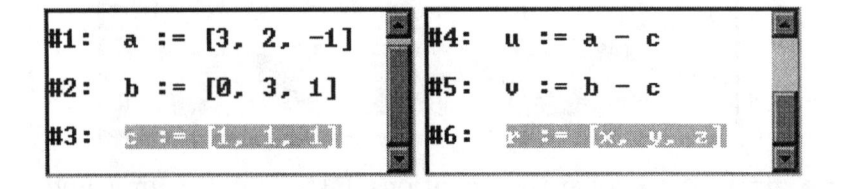

```
#1:  a := [3, 2, -1]         #4:  u := a - c
#2:  b := [0, 3, 1]          #5:  v := b - c
#3:  c := [1, 1, 1]          #6:  r := [x, y, z]
```

A suitable normal vector is then $\mathbf{n} = \mathbf{u} \times \mathbf{v}$, and the equation of the plane is $(\mathbf{r} - \mathbf{a}) \cdot \mathbf{n} = 0$. These calculations are shown below.

2.6. EQUATIONS OF PLANES

```
#7:  n := CROSS(u, v)
#8:  (r - a)·n = 0
#9:  4·x + 2·y + 5·z - 11 = 0
```

Consequently, the equation of the plane is given by

$$4x + 2y + 5z = 11.$$

Intersection of a Line and a Plane. Suppose that we wish to determine the intersection of a line and a plane given respectively by

$$\mathbf{r}(t) = \mathbf{r}_0 + t\mathbf{v} \text{ and } \langle x, y, z \rangle \cdot \mathbf{n} = \mathbf{q}_0 \cdot \mathbf{n}.$$

An example which illustrates the process in **DERIVE** is shown below.

Example 9 *Find the point of intersection of the line and plane given respectively by*

$$\mathbf{r}(t) = \langle t, 1-t, t-1/3 \rangle \text{ and } 2x - y + 3z = 3.$$

We begin be defining functions which represent our line and plane. Note that the function $\mathbf{F}(x, y, z)$ below must be set equal to zero to represent the plane.

```
#10:  F(x, y, z) := 2·x - y + 3·z - 3
#11:  R(t) := [t, 1 - t, t - 1/3]
```

Now we author an expression for the composition of $\mathbf{F}(\mathbf{r}(t)) = 0$ by using the **SUB** command and the ▣ icon.

$$f(r(t) \text{ SUB } 1, r(t) \text{ SUB } 2, r(t) \text{ SUB } 3) = 0$$

```
#12:  F((R(t))₁, (R(t))₂, (R(t))₃) = 0
#13:  6·t - 5 = 0
```

Finally, we use the **SOLVE** command to determine the value of t. This value is substituted into the function $\mathbf{R}(t)$ to find the point of intersection.

```
#14: SOLVE(6·t - 5 = 0, t)

#15: [t = 5/6]
```

```
#16: R(5/6) = [5/6, 1/6, 1/2]
```

Intersection of Two Planes. Unless two planes are parallel, their intersection will always be a line.

Example 10 *Determine the intersection of the two planes given by*

$$2x - 3y + z = 1 \text{ and } x + y + 2z = 0.$$

There are various ways to approach this problem. One is to observe the following:

- The normal vectors, $\langle 2, -3, 1 \rangle$ and $\langle 1, 1, 2 \rangle$, are not parallel; therefore the planes are not parallel; therefore the intersection of the two planes must be a line.

- Any direction vector of the line of intersection must be orthogonal to the normal vector of each of the two planes.

Another method is to take advantage of the power of **DERIVE**. We start by using the **SOLVE** command to determine x, y and z in terms of an artitrary parameter named @1.

```
#18: SOLVE([2·x - 3·y + z = 1, x + y + 2·z = 0], [x, y, z])

#19: x = @1   y = (3·@1 - 2)/7   z = (1 - 5·@1)/7
```

Then we improve the appearance of the expression in #19 by using the `Sub` icon.

2.6. EQUATIONS OF PLANES

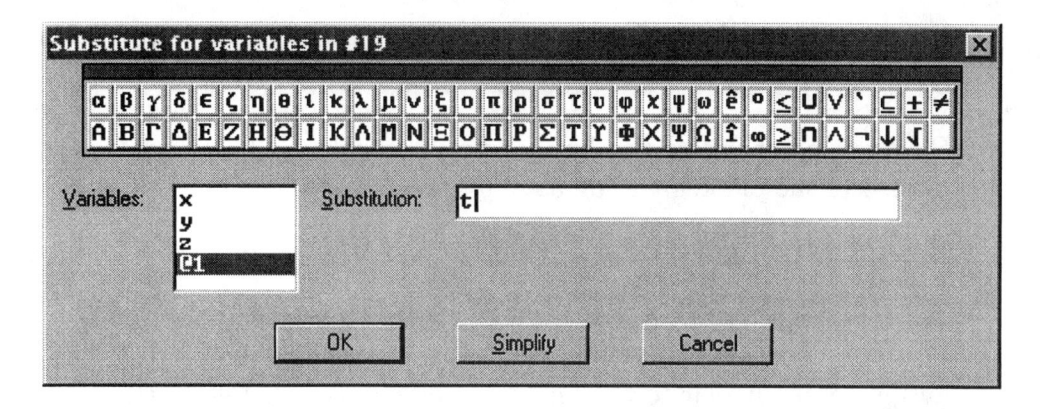

The resulting line of intersection is given by the parameterization below.

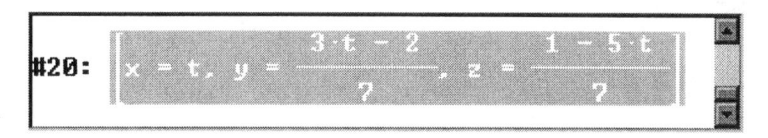

#20: $x = t, \quad y = \dfrac{3 \cdot t - 2}{?}, \quad z = \dfrac{1 - 5 \cdot t}{?}$

Exercises:

1. Find a unit vector that is perpendicular to each of the vectors $\mathbf{a} = \langle 1, 2, 3 \rangle$ and $\mathbf{b} = \langle 3, 2, 1 \rangle$.

2. Find a unit vector that is perpendicular to the plane that passes through the points $(1, 1, 1)$, $(1, 2, 2)$, and $(3, -1, 0)$.

3. Find the area of the triangle whose vertices are the points $(1, 2, 3)$, $(3, 2, 1)$, and $(2, 1, 3)$.

4. Determine the intersection of the following pairs of lines. Describe each result geometrically.

    (a) $\mathbf{r} = \langle 1 - t, 1 + 2t, t \rangle$ and $\mathbf{r} = \langle 2 - 3t, 5t - 1/2, 1 - t \rangle$
    (b) $\mathbf{r} = \langle 1 - t, 1 + 2t, t \rangle$ and $\mathbf{r} = \langle 2 - 3t, 5t - 1/2, 1 + t \rangle$
    (c) $\mathbf{r} = \langle 1 - t, 1 + 2t, t \rangle$ and $\mathbf{r} = \langle t - 1, 5 - 2t, 2 - t \rangle$

5. Find a parametric equation for the line that passes through the point $(3, -2, 1)$ and is perpendicular to each of the lines $\mathbf{r} = \langle 1 - t, 1 + 2t, t \rangle$ and $\mathbf{r} = \langle 2 - 3t, 5t - 1/2, 1 - t \rangle$.

6. Find the point of intersection of the line and plane given respectively by $\langle x, y, z \rangle = \langle 2t, 1-t, t \rangle$ and $3x - 2y + z = 5$.

7. Find the point where the line passing through $(5, 3, 1)$ and $(1, 1, 2)$ intersects the plane given by $3x + y + z = 3$.

8. Determine the intersection of the two planes given by $x - y + z = 2$ and $2x + y - 2z = 0$.

9. Find the equation of the plane that passes through the point $(3, 2, 1)$ and is perpendicular to the line given by $\mathbf{r} = \langle 1-t, 1+2t, t \rangle$.

Chapter 3

Surfaces

This chapter introduces a number of capabilities of the **DERIVE** for visualizing the graph of a function of two variables.

- Related material is found in Chapter 9 of Stewart's **Multivariable Calculus: Concepts and Contexts**.

3.1 Graphs of Functions of Two Variables

- Graphs of functions of two variables are discussed in Section 9.6 of Stewart's text.

To plot the graph of a function of two variables with **DERIVE**, we must first create the function in the **Algebra** window. Once that is done, we press the **3-D Plot** icon and select **Plot!**. It is not unusual for this to result in a pot which looks horrible. That's because we have to be careful when we create a **3-D Plot**. It's important to make good choices for the lengths of the axes, the viewpoint and the grid size. The most important of these is usually the lengths of the axes. We demonstrate this in the example below.

Example 11 *Let's plot the graph of $z = \sin(x + 2y)$.*

We start by creating the expression in the **Algebra** window. Then we click the **3-D Plot** icon and select **Plot!**.

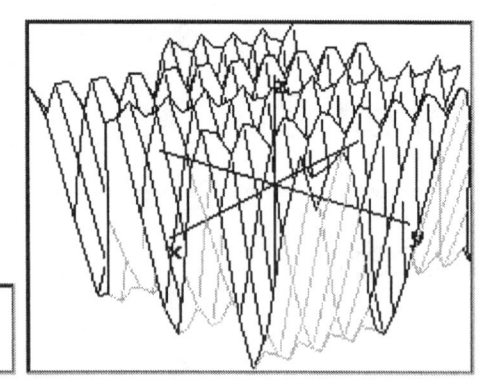

Notice that the graph looks terrible. That's because we did not consider the length of our axes. If we select **Set** and click on **Length** then we can change the length of the axis as shown below to create a better plot.

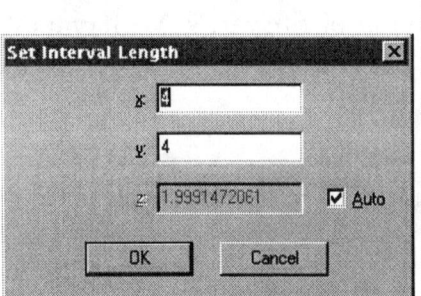

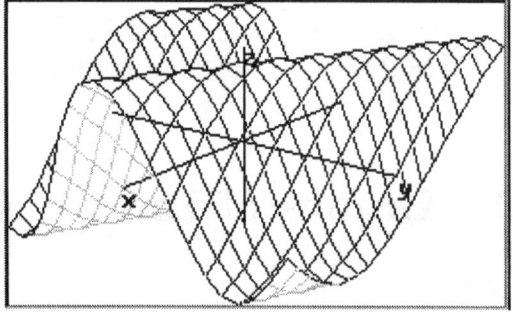

The graph above is sketched with the viewpoint (or **Eye**) Set at $x = 8$, $y = 6$ and z set to **auto**. In the sketch above, **DERIVE** used the value $z = 3$. Perhaps a better view of the surface will result from a "higher" viewpoint. The result of using $z = 8$ is shown below.

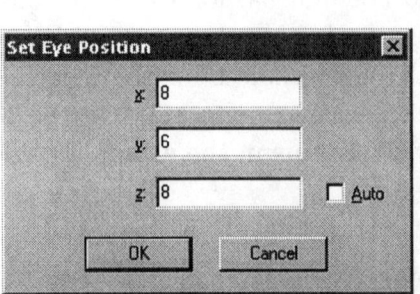

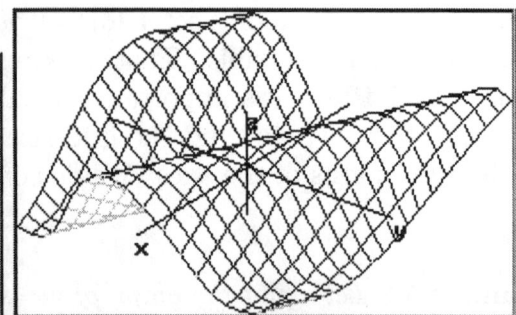

3.1. GRAPHS OF FUNCTIONS OF TWO VARIABLES 39

The grid size in the sketch above is **Set** to 20 in each of the x and y directions. Let's now see what happens when we double the grid size.

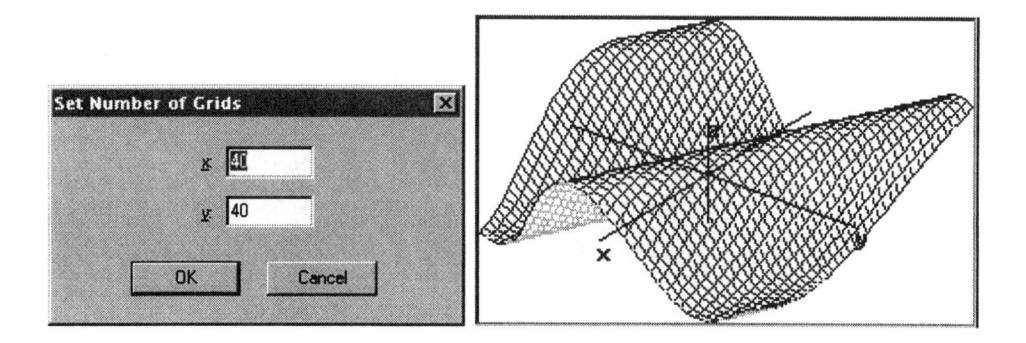

We now have a rather nice plot of the surface.

Remark: The actual range for the values of x and y are determined by **Set**ting both the **Length** and the **Center**. The plots above were created with $(0, 0, 0)$ as the **Center**.

Example 12 *Now that we have explored a few of the **3-D Plot** settings available to us, let's plot a different surface. Consider the graph of*

$$z = 3x - x^3 - 2xy^2 + 2y.$$

The following two screens show the set-up for the **Length** of the axis and the **Eye** position.

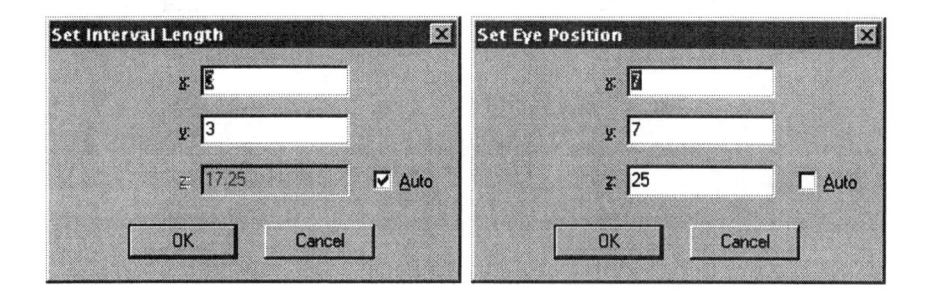

Finally, we enter the expression in the **Algebra** window, click on the **3-D Plot** icon and select **Plot!**.

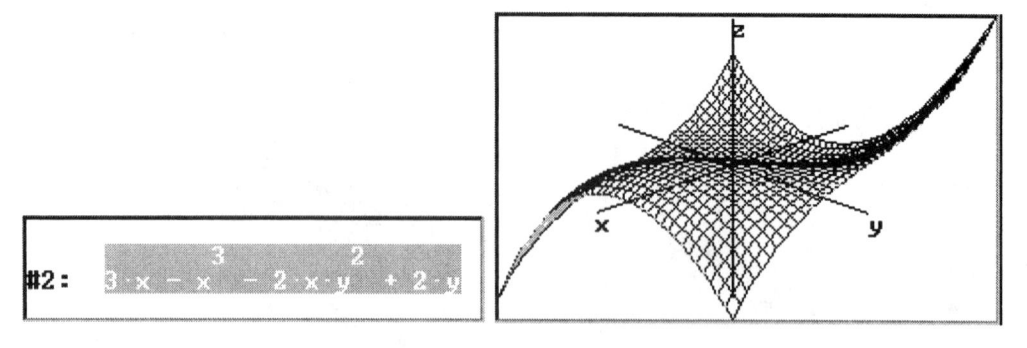

#2: $3 \cdot x^3 - x^2 - 2 \cdot x \cdot y + 2 \cdot y$

3.2 Traces and Contours

- Traces and contours are discussed in Section 9.6 of Stewart's **Multivariable Calculus: Concepts and Contexts**. However, the term "contour" is not used until Chapter 11.

To understand the graph of a equation in three variables x, y, and z, it is often helpful to look at cross-sections obtained by setting one of the three variables equal to a constant and graphing the resulting equation in the other two variables. Such cross-sections are examples of *traces*.

Example 13 *To illustrate traces, let's consider the equation*

$$x^2 + y^3 - z = 0.$$

Note that the graph of this equation is the same as the graph of the function $f(x,y) = x^2 + y^3$. To create the graph of the function, we set the **Length** and **Eye** as shown below.

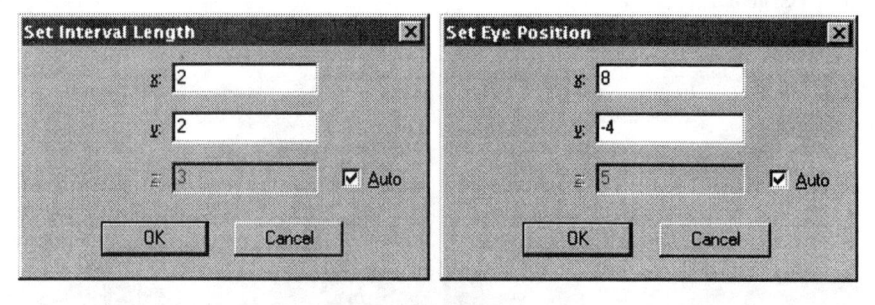

Then we define the function in the **Algebra** window

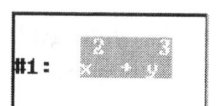

3.2. TRACES AND CONTOURS

#1: $x^2 + y^3$

and create the plot.

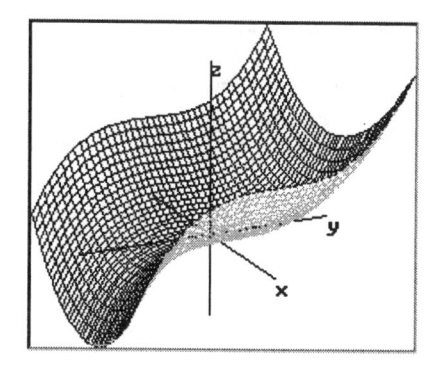

Setting x to a constant amounts to slicing the surface vertically, parallel to the yz-plane. Traces in $x = k$ are the cubic curves $z = k^2 + y^3$. To plot a sampling of these curves we create a **VECTOR** of expressions by authoring the expression

$$\text{VECTOR}(k\hat{\,}2 + y\hat{\,}3, k, 0, 2, 1)$$

and selecting the ▨ icon. This gives the **Algebra Window** shown below.

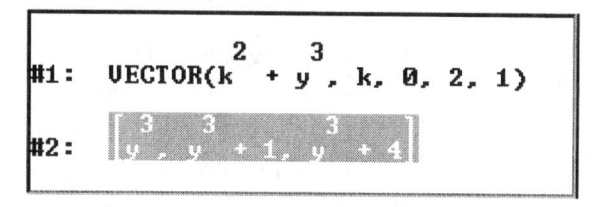

#1: VECTOR($k^2 + y^3$, k, 0, 2, 1)

#2: $[y^3, y^3 + 1, y^3 + 4]$

Finally, we click on the **2-D Plot** icon, set the **Range** of the axes and create our plot.

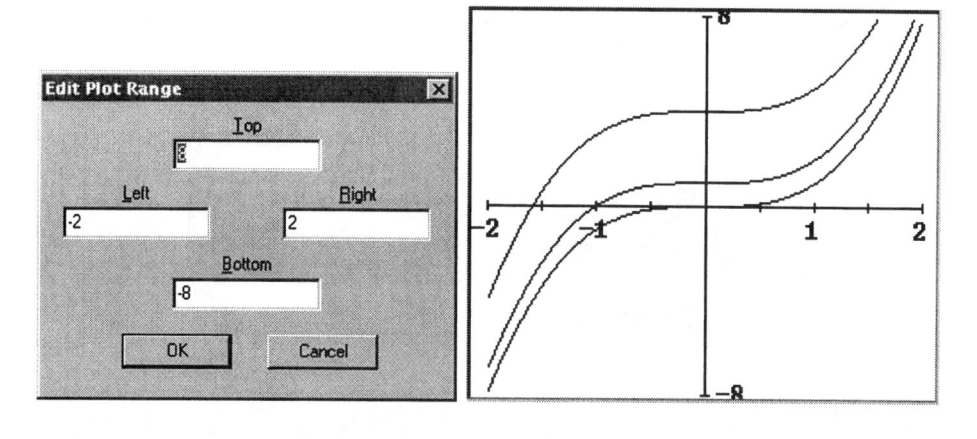

Setting y to a constant amounts to slicing the surface vertically, parallel to the xz-plane. Traces in $y = k$ are the parabolic curves $z = x^2 + k^3$. We can proceed as above to create the following windows.

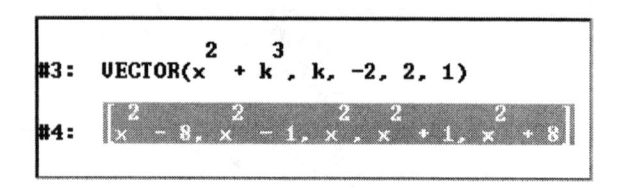

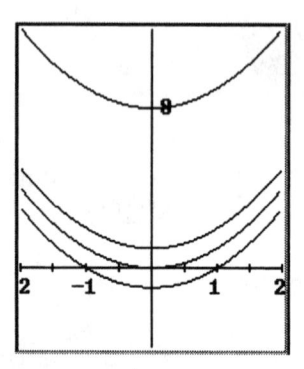

#3: VECTOR($x^2 + k^3$, k, -2, 2, 1)

#4: $[x^2 - 8, x^2 - 1, x^2, x^2 + 1, x^2 + 8]$

Setting z to a constant amounts to slicing the surface horizontally, parallel to the xy-plane. The traces in $z = k$ are the curves $x^2 + y^3 = k$. These curves can be created in one of two different ways. One method is to solve for y in the form $y = (k - x^3)^{1/3}$ and then plot the graph of this expression for various values of k. This method is actually harder than you might think (try it!!), and since it is not always possible to solve for the variable y, we show an alternate method. Here we simply form the equations $x^2 + y^3 = k$ for various values of k and ask for a plot. This requires some adjustment of the **grid size**.

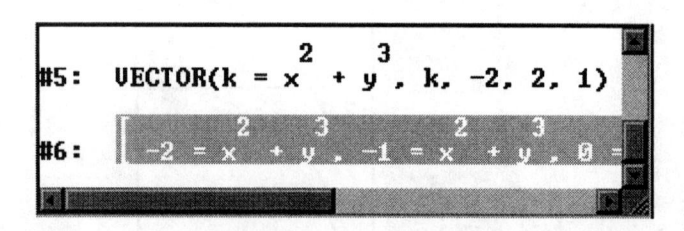

#5: VECTOR($k = x^2 + y^3$, k, -2, 2, 1)

#6: $[-2 = x^2 + y^3, -1 = x^2 + y^3, 0 = \ldots]$

3.2. TRACES AND CONTOURS

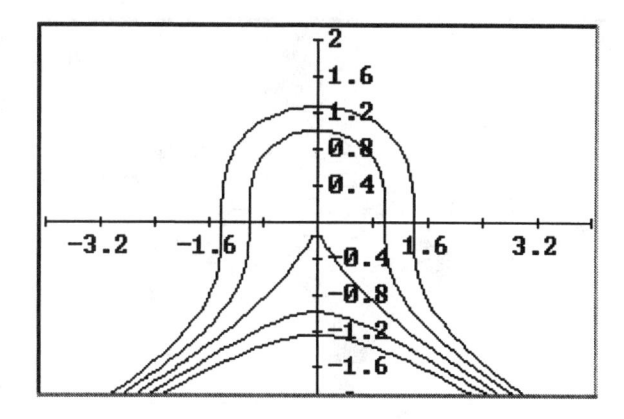

Horizontal traces of the graph of a function f are especially important and are usually called **contours**, or **level curves**, of the function f.

Example 14 *To illustrate the creation of level curves once more, let's look at the function*

$$f(x,y) = 3x - x^3 - 2xy^2 + 2y.$$

A plot of the surface is shown on the right below.

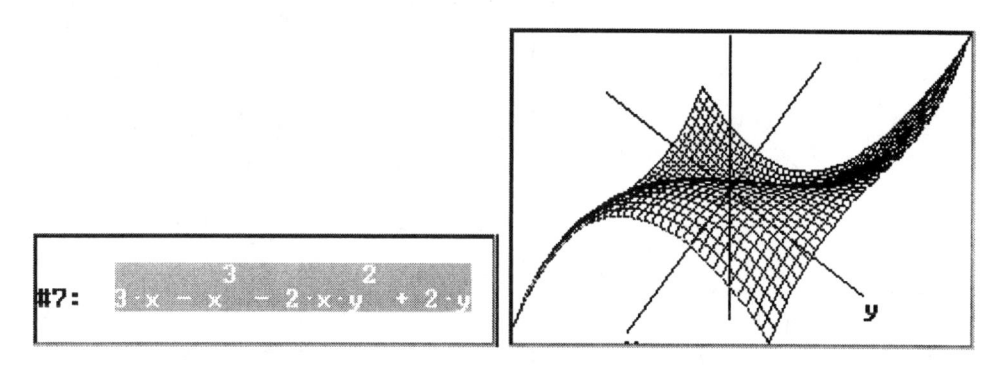

#7: $3 \cdot x - x^3 - 2 \cdot x \cdot y^2 + 2 \cdot y$

The plot above was created with the **Eye** at $x = 4$, $y = 3$ and $z = 15$. Level curves can be obtained as shown below.

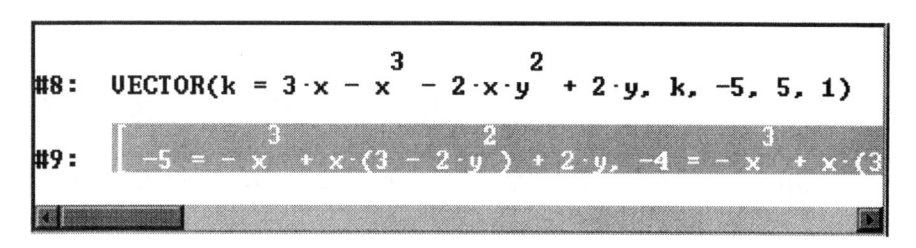

#8: VECTOR(k = 3·x - x³ - 2·x·y² + 2·y, k, -5, 5, 1)

#9: $-5 = -x^3 + x \cdot (3 - 2 \cdot y^2) + 2 \cdot y, \; -4 = -x^3 + x \cdot (3$

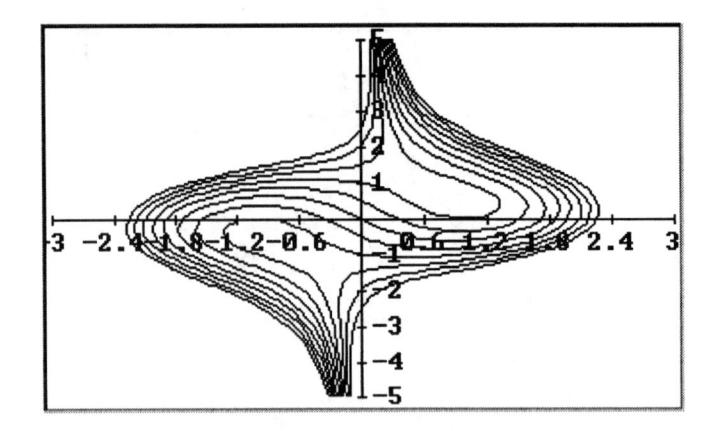

(Note that increasing **grid size** produces smoother curves.)

Example 15 *From the example above, we can see that it is a simple matter to create implicit plots with* **DERIVE**. *Create the implicit plot of* $f(x,y) = 0$ *where*
$$f(x,y) = x^3 - 2xy + y^3.$$

The plot is shown below.

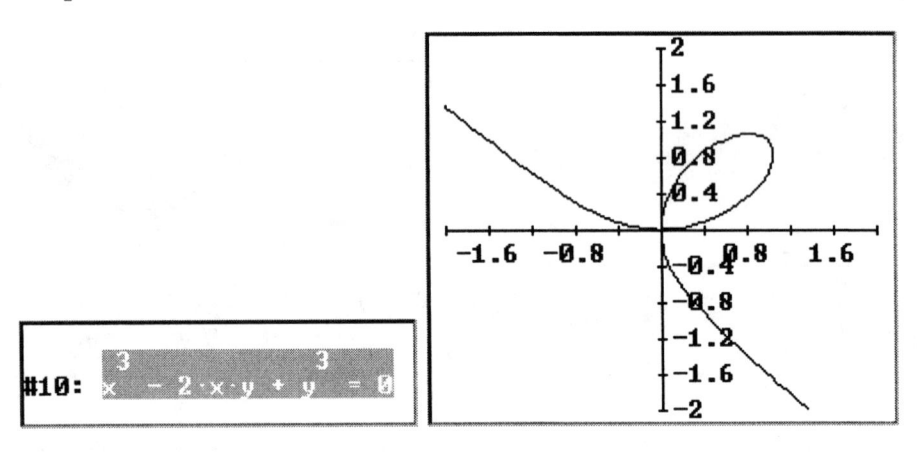

Exercises:

1. Let $f(x,y) = xy$.

    (a) Obtain a good **3-D Plot** of the graph of f by choosing appropriate window variables.

3.2. TRACES AND CONTOURS

    (b) Plot families of traces in $x = k$ and $y = k$.

    (c) Create a level curve plot of f.

    (d) Create an implicit plot of $f(x, y) = 0$.

2. Repeat exercise 1 for each of the following functions:

    (a) $f(x, y) = x y^2$.

    (b) $f(x, y) = \sin(2x + y) + \cos(x - y)$

    (c) $f(x, y) = \cos(x + \sin y)$

    (d) $f(x, y) = \cos(x y) - \cos x \cos y$

Chapter 4

Vector-valued Functions

In this chapter we explore curves in two and three dimensional space in **DERIVE**.

- Related material can be found in Chapter 10 of Stewart's **Multivariable Calculus: Concepts and Contexts**.

4.1 Space Curves

- Vector functions and space curves are the subject of Chapter 10 in Stewart's text.

DERIVE has a built in facility for plotting parametric curves in two-dimensions, and there is a utility that can be loaded to help plot parametric curves in three dimensions (space curves).

Example 16 *Let's plot the curve given by*

$$x = \sin\left(t + \frac{\pi}{4}\right), \; y = \cos(t)$$

along with a sampling of position vectors along the curve.

We start by defining a vector valued function $\mathbf{v}(t) = [\sin(t + \pi/4), \cos(t)]$ and plotting it in a **2-D Plot** window.

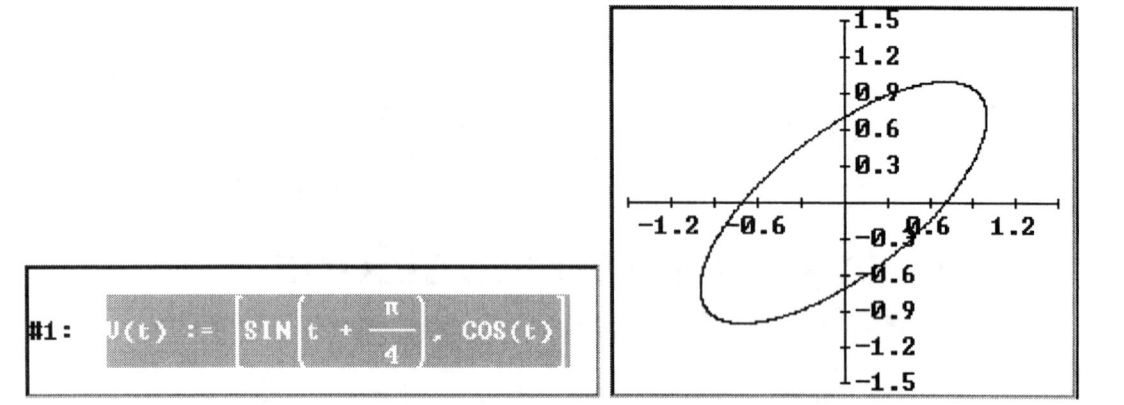

#1: $v(t) := \left[\text{SIN}\left(t + \dfrac{\pi}{4}\right), \text{COS}(t) \right]$

Here we have set the parameter range (for t) from $-\pi$ to π. We can plot a number of position vectors associated with this curve by using the **VECTOR** command. We begin by issuing the command shown on the left below, followed by pressing the = icon.

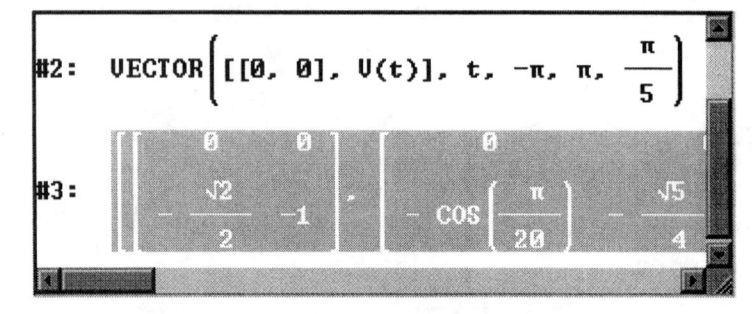

#2: $\text{VECTOR}\left([[0, 0], v(t)], t, -\pi, \pi, \dfrac{\pi}{5} \right)$

#3: $\left[\left[0, 0 \right], \left[-\dfrac{\sqrt{2}}{2}, -1 \right] \right], \left[0, -\cos\left(\dfrac{\pi}{20}\right), -\dfrac{\sqrt{5}}{4} \right]$

Then we create the **2-D Plot** shown below.

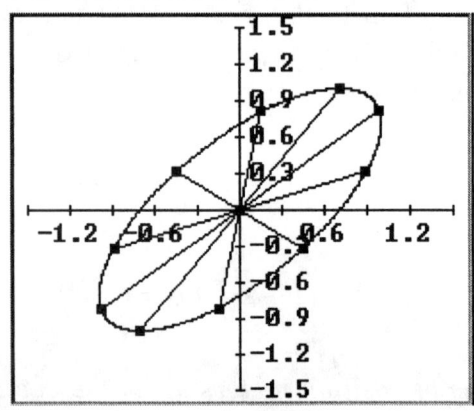

4.1. SPACE CURVES

Space Curves. Our main concern here will be functions of one variable whose values are three-dimensional vectors:

$$\mathbf{r}(t) = \langle f(t), g(t), h(t) \rangle = f(t)\mathbf{i} + g(t)\mathbf{j} + h(t)\mathbf{k}$$

The set of points $(f(t), g(t), h(t))$ where t varies over the domain of $\mathbf{r}(t)$, is called a **space curve**. In other words, the space curve defined by $\mathbf{r}(t)$ is the set of points described by the parametric equations

$$x = f(t), \quad y = g(t), \quad z = h(t).$$

Likewise, we say that the function $\mathbf{r}(t)$ is a **parametrization** of the space curve.

Unfortunately, **DERIVE** does not have built-in capability for plotting space curves. However, there is a utility file which can be loaded so that space curves can be plotted. We demonstrate the use of this utility file in the examples below. *Project 8.5 asks for an improvement of the feature that we demonstrate here.*

Example 17 *Create the plot of the helix parametrized by*

$$x = \sin(t), \quad y = \cos(t), \quad z = t/10,$$

for $0 \leq t \leq 20$.

We start by clicking on **File**, selecting **Load**, and clicking on **Utility**. Then we open the **Graphics.mth** file. This allows us to use the **AXES** and **ISOMETRIC** commands. The **axes** command plots the x, y and z axes, and the **ISOMETRIC** command plots the curve in 3-space. Since the plot is actually given in a **2-D Plot** window, it is important to turn off the plotting of the axes in this window. This is done from the **2-D Plot** window by clicking on **Options** and then **Axes**.

We demonstrate the process below. We begin by entering a vector function representing the parameterization above. Then we use the **ISOMETRIC** command and evaluate the result by clicking on the = icon.

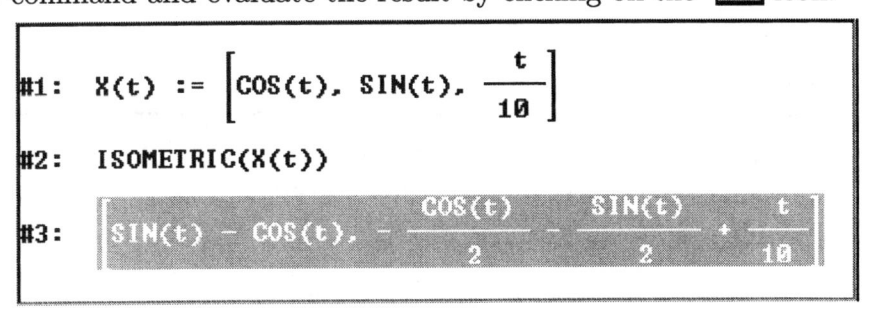

```
#1:   X(t) := [COS(t), SIN(t), t/10]

#2:   ISOMETRIC(X(t))

#3:   [SIN(t) - COS(t), -COS(t)/2 - SIN(t)/2 + t/10]
```

This projects the parametric curve onto a view plane in 2-dimensional space. We plot this curve in a **2-D Plot** window below. Notice that the axes have been turned off (as stated above).

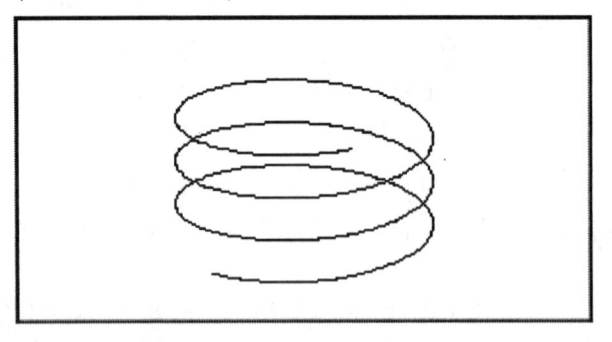

We place coordinate axes in this plot by using the **axes** command. The lengths of the coordinate axes are determined by supplying numerical values to a window prompt.

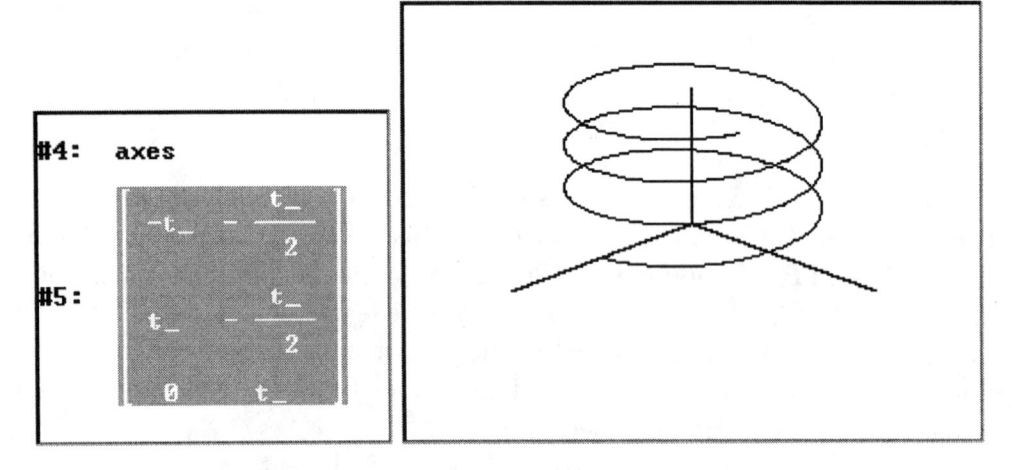

```
#4:    axes
                    t_
           -t_  -  ──
                    2
#5:        t_      t_
           t_   -  ──
                    2
           0       t_
```

Example 18 *Plot of the curve parametrized by*

$$x = 2\sin(t), \quad y = 2\cos(t), \quad z = \sin(5t),$$

for $0 \leq t \leq 2\pi$.

The solution is shown below.

```
#6:   X(t) := [2·SIN(t), 2·COS(t), SIN(5·t)]
#7:   ISOMETRIC(X(t))
#8:   [2·COS(t) - 2·SIN(t), SIN(5·t) - COS(t) - SIN(t)]
```

4.1. SPACE CURVES

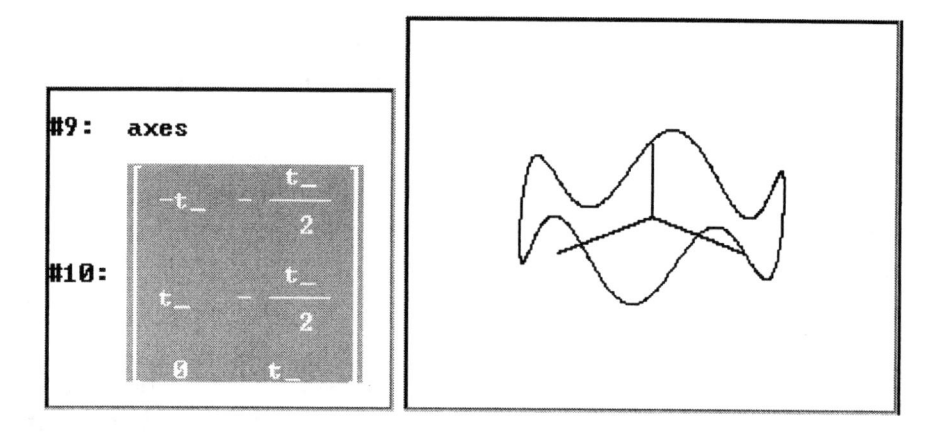

Exercises: For each of the plane curves in exercises 1–5, follow Example 1 to plot the curve along with a sampling of position vectors.

1. $\mathbf{r}(t) = \langle \cos(t), \sin(t) \rangle$
2. $\mathbf{r}(t) = \langle \cos(t/2), \sin(t) \rangle$
3. $\mathbf{r}(t) = \langle t, \cos(t) \rangle, \quad 0 \le t \le 2\pi$
4. $\mathbf{r}(t) = \langle e^{-t/2} \cos(t), e^{-t/2} \sin(t) \rangle, \quad 0 \le t \le 2\pi$
5. $\mathbf{r}(t) = \langle 2\sin(t/5) + \sin(t), 2\cos(t/5) + \cos(t) \rangle$

Plot each of the space curves in exercises 6–8.

6. $\mathbf{r}(t) = \langle \cos(t), \sin(t), 2t^2/(3+t^2) \rangle, \quad 0 \le t \le 4\pi$
7. $\mathbf{r}(t) = \langle \cos(t/2), \sin(t), t/6 \rangle, \quad 0 \le t \le 2\pi$
8. $\mathbf{r}(t) = \langle e^{-t} \cos(t), e^{-t} \sin(t), t/6 \rangle, \quad 0 \le t \le 4\pi$

For each of the pairs of surfaces in exercises 9 and 10, find and plot a parametrization of the intersection.

9. $z = 4 - x^2 - y^2$ and $z = x^2 + 2y^2 - xy$
10. $z = 2x^2 + 2y^2 - 3$ and $z = x^2 - y^2$

4.2 Derivatives

- For development of the derivative of a vector-valued function and its properties, see Section 10.2 of Stewart's **Multivariable Calculus: Concepts and Contexts.**

We know from Stewart's text that the derivative of a vector-valued function is defined in a manner analogous to that of an ordinary real-valued function.

$$\mathbf{r}'(t) = \lim_{h \to 0} \frac{\mathbf{r}(t+h) - \mathbf{r}(t)}{h}$$

The difference quotient on the right side is a vector quantity; hence so is its limit as $h \to 0$. In fact, $\mathbf{r}'(t_0)$ is a vector that is tangent to the curve at the point whose position vector is $\mathbf{r}(t_0)$, provided that it exists and is not $\vec{0}$.

Example 19 *To illustrate the limiting process that defines $\mathbf{r}'(t_0)$, let's plot the curve parametrized by*

$$\mathbf{r}(t) = \langle \cos(t), \sin(3t) \rangle$$

together with a few vectors of the form

$$\frac{\mathbf{r}(\pi/6 + h) - \mathbf{r}(\pi/6)}{h}$$

with small values of h.

We will plot these vectors with a common initial point whose position vector is

$$\mathbf{r}(\pi/6) = \langle \cos(\pi/6), \sin(3\pi/6) \rangle = \left\langle \sqrt{3}/2, 1 \right\rangle.$$

So the first thing we need to do is enter definitions of **r(t)** and **dr(t,h)**.

```
#1:   R(t) := [COS(t), SIN(3·t)]

#2:   DR(t, h) := R(t + h) - R(t)
                 ─────────────────
                         h
```

Then we set up the matrix needed to plot the tangent vectors using the **VECTOR** command.

4.2. DERIVATIVES

```
VECTOR([R(π/6), R(π/6) + DR(π/6, h)], h, 0.05, 0.35, 0.1)
```

#3: $\text{VECTOR}\left(\left[R\left(\dfrac{\pi}{6}\right),\ R\left(\dfrac{\pi}{6}\right) + DR\left(\dfrac{\pi}{6}, h\right)\right],\ h,\ 0.05,\ 0.35,\ 0.1\right)$

#4: $\begin{bmatrix} 0.866025 & 1 \\ 0.344582 & 0.775420 \end{bmatrix}$, $\begin{bmatrix} 0.866025 & 1 \\ 0.303067 & 0.336313 \end{bmatrix}$, $\begin{bmatrix} 0.8660\ldots \\ 0.26352\ldots \end{bmatrix}$

Finally, we plot the expressions which appear in #1 and #4 above.

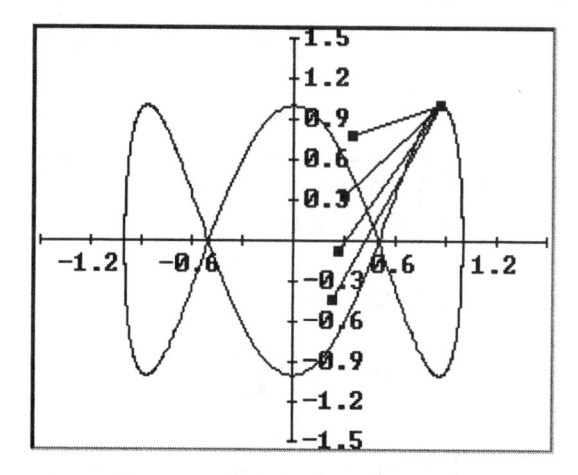

So we see that as $h \to 0$ the vectors given by the difference quotient appear to approach a vector that is tangent to the curve at the point whose position vector is $\mathbf{r}(t_0)$.

Example 20 *Using the same function as in Example 4, let's plot a portion of the curve along with approximations to the derivative at several points along the curve. We'll compute the approximations by evaluating the difference quotient with $h = .01$.*

We use the functions $\mathbf{r}(t)$ and $\mathbf{dr}(t, h)$ defined above to create the matrix needed to create the tangent vectors near $\pi/6$, $\pi/3$, $\pi/2$, $2\pi/3$, $5\pi/6$ and 2π.

VECTOR$([\mathbf{r}(n*\mathbf{pi}/6), \mathbf{r}(n*\mathbf{pi}/6) + \mathbf{dr}(n*\mathbf{pi}/6, 0.01)], n, 1, 5, 1)$

This yields

54 CHAPTER 4. VECTOR-VALUED FUNCTIONS

#6: $\text{VECTOR}\left(\left[R\left(\dfrac{n\cdot\pi}{6}\right), R\left(\dfrac{n\cdot\pi}{6}\right) + DR\left(\dfrac{n\cdot\pi}{6}, 0.01\right)\right], n, 1, 5, 1\right)$

#7: $\left[\begin{bmatrix} 0.866025 & 1 \\ 0.361689 & 0.955005 \end{bmatrix}, \begin{bmatrix} 0.5 & 0 \\ -0.368514 & -2.99960 \end{bmatrix}, \begin{bmatrix} 0 & 0 \\ -1 & -0. \end{bmatrix}\right.$

Then plotting #1 and #7 gives the following plot.

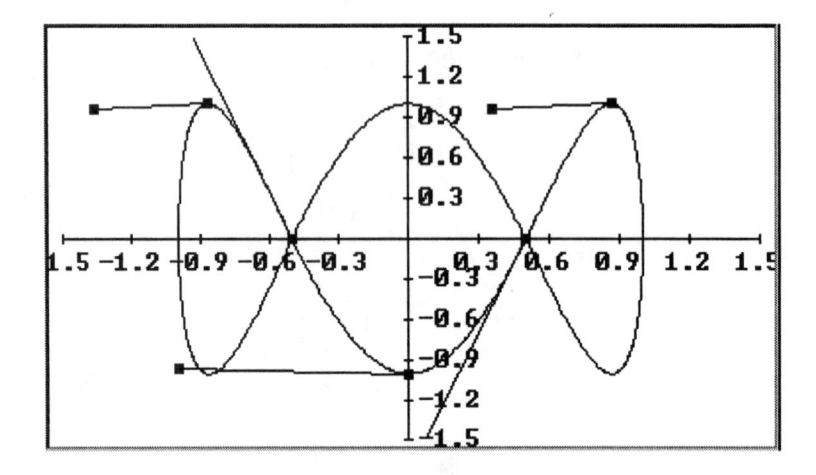

Differentiation. Differentiation of vector-valued functions is done term by term, and is handled automatically by **DERIVE**'s **DIF** operator, or by the "prime" operator.

$\text{DIF}(r(t), t) =$

#1: $R(t) := [\cos(t), \sin(3\cdot t)]$

#2: $\dfrac{d}{dt} R(t) = [-\sin(t), 3\cos(3\cdot t)]$

#3: $R'(t) = [-\sin(t), 3\cos(3\cdot t)]$

Until we start working with functions of several variables, the "prime" operator is the easiest to work with.

4.2. DERIVATIVES

Example 21 *Plot the plane curve given by*

$$r(t) = \langle \cos(t + .25), \sin(2t) \rangle$$

along with the derivative as a tangent vector to curve at each of $t = k\pi/6$, $k = 0, \ldots, 11$.

The first thing we do is define $\mathbf{r}(t)$. Then we use the **VECTOR** command to create a matrix which will yield the desired tangent vectors.

```
vector( [r(k*pi/6),r(k*pi/6)+r'(k*pi/6)],k,0,11,1)
```

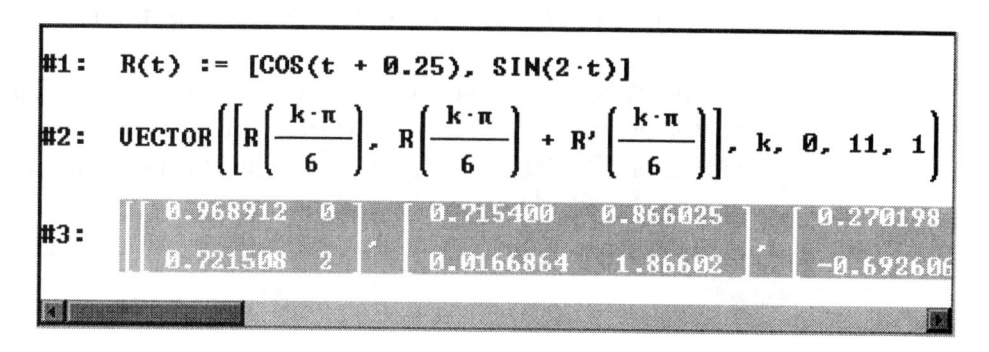

#1: R(t) := [COS(t + 0.25), SIN(2·t)]

#2: VECTOR$\left(\left[R\left(\frac{k \cdot \pi}{6}\right), R\left(\frac{k \cdot \pi}{6}\right) + R'\left(\frac{k \cdot \pi}{6}\right)\right], k, 0, 11, 1\right)$

#3: $\begin{bmatrix} 0.968912 & 0 \\ 0.721508 & 2 \end{bmatrix}$, $\begin{bmatrix} 0.715400 & 0.866025 \\ 0.0166864 & 1.86602 \end{bmatrix}$, $\begin{bmatrix} 0.270198 \\ -0.692606 \end{bmatrix}$

Finally, we plot #1 and #3 in a **2-D Plot** window.

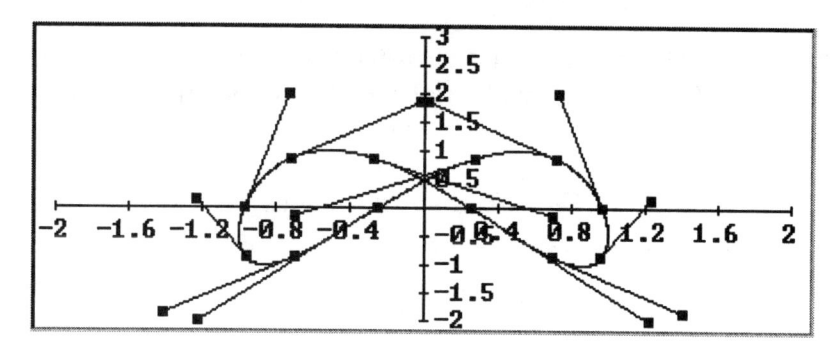

Exercises:

1-5. For each of the plane curves in Exercises 1-5 from Section 4.1, plot the curve along with a sampling of derivative vectors.

6-8. For each of the space curves in Exercises 6-8 from Section 4.1, plot the curve along with a sampling of derivative vectors.

9. Let C be the space curve parametrized by
$$\mathbf{r}(t) = 4\left\langle \sin^2(t),\ \cos^2(3t),\ \sin(t) \right\rangle.$$

   (a) Find a value of t such that $\mathbf{r}(t) = \langle 3, 3, 1 \rangle$.
   (b) Find a unit vector that is tangent to C at $(3, 3, 1)$.
   (c) Find the equation of the plane that passes through $(3, 3, 1)$ and is orthogonal to C at that point.

10. Let C_r and C_q be the two space curves parametrized respectively by
$$\mathbf{r}(t) = \langle \sin(2\pi t),\ \cos^2(\pi t),\ 2t \rangle \quad \text{and} \quad \mathbf{q}(t) = \langle 2t,\ 2 - 3t,\ 1 - t \rangle.$$

   (a) Find the point at which C_r and C_q intersect. (Suggestion: The position vector of this point may be attained by $\mathbf{r}(t)$ and $\mathbf{q}(t)$ at different values of t; so look at $\mathbf{r}(s) = \mathbf{q}(t)$. The third coordinates give an easy relationship between s and t.)
   (b) Find the (acute) angle between C_r and C_q at the point where they intersect.

4.3 Arc Length and Curvature

- The notions of arc length and curvature of a space curve are developed in Section 10.3 of Stewart's **Multivariable Calculus: Concepts and Contexts**.

We know from Stewart's text that the length of a space curve with a smooth parametrization
$$\mathbf{r}(t) = \langle f(t), g(t), h(t) \rangle, \quad a \le t \le b$$
is the integral of the length of the derivative over the interval $a \le t \le b$:
$$L = \int_a^b |\mathbf{r}'(t)|\,dt = \int_a^b \sqrt{f'(t)^2 + g'(t)^2 + h'(t)^2}\,dt.$$

Example 22 *Compute the length of the space curve parametrized by*
$$\mathbf{r}(t) = \langle \cos(e^t),\ \sin(e^t),\ e^t/10 \rangle, \quad 0 \le t \le 2.$$

4.3. ARC LENGTH AND CURVATURE

It is easy to see that $\mathbf{r}(t)$ is just a peculiar parametrization of a simple helix with radius 1. To compute the arc length, we integrate the length of the derivative over the appropriate interval.

```
int(abs(r'(t)),t,0,2)=
```

#1: $R(t) := \left[\cos(\text{EXP}(t)), \sin(\text{EXP}(t)), \dfrac{\text{EXP}(t)}{10} \right]$

#2: $\displaystyle\int_0^2 |R'(t)|\,dt = \dfrac{\sqrt{101}\cdot e^2}{10} - \dfrac{\sqrt{101}}{10}$

So the exact value of the arc length is $(e^2 - 1)\sqrt{1.01}$. The approximate value is 6.42092.

Example 23 *Plot and compute the length of the space curve parameterized by*
$$\mathbf{r}(t) = \langle \sin(t), \cos(t), \sin(2t) \rangle, \quad 0 \le t \le \pi.$$

We compute the length of the space curve by following the example above.

#1: $R(t) := [\sin(t), \cos(t), \sin(2 \cdot t)]$

#2: $\displaystyle\int_0^\pi |R'(t)|\,dt$

However, in this case **DERIVE** is unable to compute the exact integral. This is because computation of an exact value is not possible except in terms of a special, non-elementary function known as an *elliptic integral*. Consequently, we just simplify the integral and ask for an approximation.

#3: $\displaystyle\int_0^\pi \sqrt{2\cdot\cos(4\cdot t) + 3}\,dt$

#4: 5.27036

The curve is plotted by loading the **Graphics.mth** utility file and using the **ISOMETRIC** command to create a plot in a **2-D Plot** window. We have turned the **axes** off below and used the **axes** command to obtain the plot of the x, y and z axes.

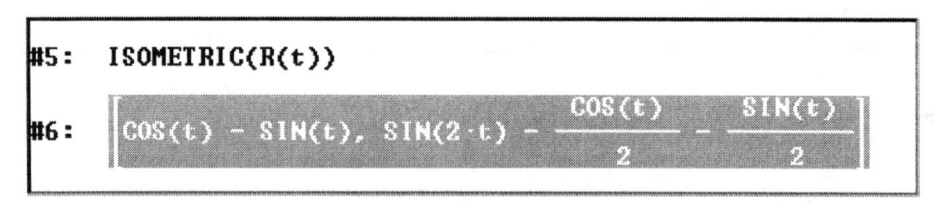

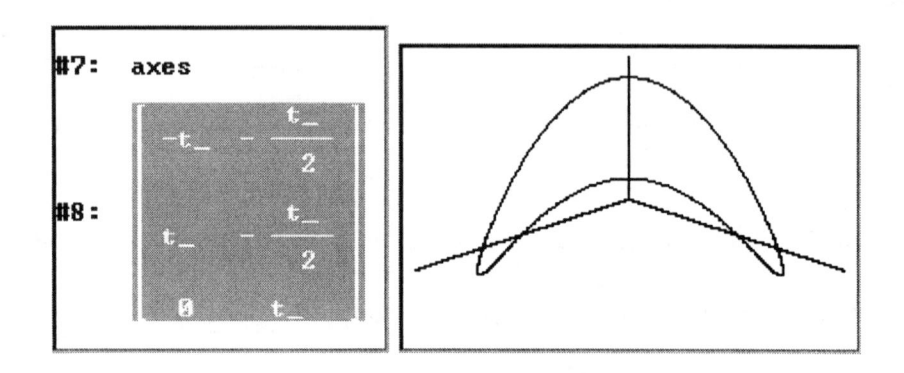

```
#5:   ISOMETRIC(R(t))

#6:   COS(t) - SIN(t), SIN(2·t) - COS(t)/2 - SIN(t)/2
```

```
#7:   axes

#8:   ⎡ -t_   - t_/2 ⎤
      ⎢ t_    - t_/2 ⎥
      ⎣ 0       t_   ⎦
```

The Unit Tangent Vector. Suppose that **C** is a curve with smooth parametrization $\mathbf{r}(t)$ (so that **r** is differentiable with $\mathbf{r}'(t) \neq \vec{0}$). The derivative $\mathbf{r}'(t)$ provides us with a tangent vector to the curve at any given point, but the length of that vector is dependent on the parametrization. However, at each point along the curve there is a unit tangent vector **T** which is characteristic of the curve itself and not dependent on the parametrization. The unit tangent vector can be parametrized by

$$\mathbf{T}(t) = \mathbf{r}'(t)/|\mathbf{r}'(t)|.$$

Example 24 *Consider the plane curve parametrized by*

$$\mathbf{r}(t) = \langle \sin(2t) - 3\cos(t), \cos(2t) \rangle, \quad 0 \leq t \leq 2\pi.$$

Plot the curve along with several unit tangent vectors given by $\mathbf{T}(t)$.

We'll first enter the parametrization as $\mathbf{r}(t)$ and create $\mathbf{T}(t)$.

4.3. ARC LENGTH AND CURVATURE

#1:   R(t) := [SIN(2·t) - 3·COS(t), COS(2·t)]

#2:   R'(t)

#3:   [2·COS(2·t) + 3·SIN(t), - 2·SIN(2·t)]

#4:   UT(t) = $\dfrac{[2 \cdot COS(2 \cdot t) + 3 \cdot SIN(t), - 2 \cdot SIN(2 \cdot t)]}{|[2 \cdot COS(2 \cdot t) + 3 \cdot SIN(t), - 2 \cdot SIN(2 \cdot t)]|}$

Then we use the **VECTOR** command to create a matrix which can be used to plot the vectors $\mathbf{T}(t)$ along the curve.

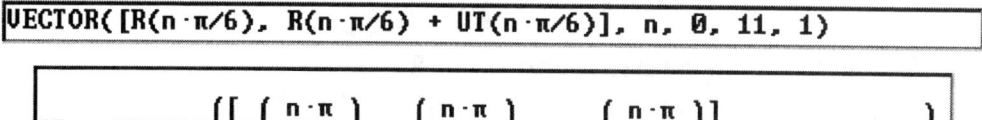

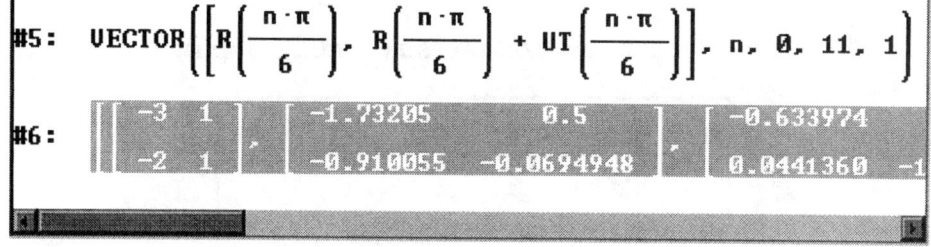

VECTOR([R(n·π/6), R(n·π/6) + UT(n·π/6)], n, 0, 11, 1)

#5:   VECTOR$\left(\left[R\left(\dfrac{n \cdot \pi}{6}\right), R\left(\dfrac{n \cdot \pi}{6}\right) + UT\left(\dfrac{n \cdot \pi}{6}\right)\right], n, 0, 11, 1\right)$

#6: $\left[\begin{array}{cc} -3 & 1 \\ -2 & 1 \end{array}\right]$, $\left[\begin{array}{cc} -1.73205 & 0.5 \\ -0.910055 & -0.0694948 \end{array}\right]$, $\left[\begin{array}{cc} -0.633974 & \\ 0.0441360 & -1 \end{array}\right]$

As in the earlier examples, we have used the values $t = k\pi/6$ for $k = 0, ..., 11$. The plot is shown below.

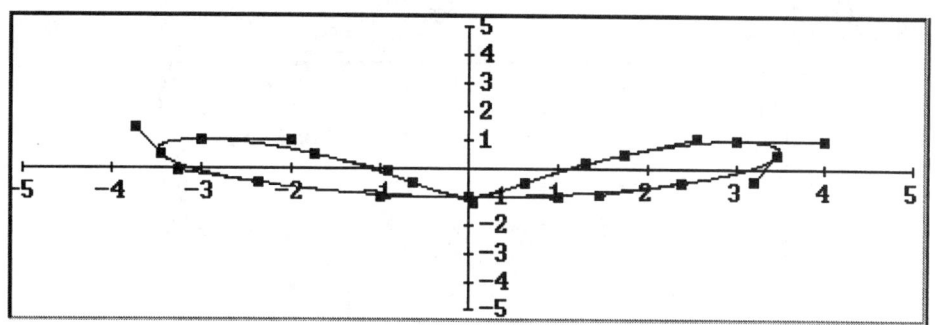

Curvature. Since the unit tangent vector $\mathbf{T}(t)$ has constant length, only its direction changes as t varies. The derivative of T with respect to arc length provides a parametrization-independent indication of the way in which the direction of the curve is changing at any given point on the curve. In fact, the **curvature** of a curve is defined by

$$\kappa = \left|\dfrac{d\mathbf{T}}{ds}\right|.$$

Application of the Chain Rule (see Section 10.2 of **Multivariable Calculus: Concepts and Contexts**) provides a more convenient formula for κ that does not require parametrization by arc length. Given any smooth parametrization, the curvature at the point whose position vector is $\mathbf{r}(t)$ is given by

$$\kappa(t) = \frac{|\mathbf{T}'(t)|}{|\mathbf{r}'(t)|}.$$

Example 25 *Compute the curvature of the ellipse parametrized by*

$$\mathbf{r}(t) = \langle 3\cos(t), \sin(t) \rangle.$$

Plot the graph of the curvature for $0 \leq t \leq 2\pi$.

Let's first enter the parametrization as **r(t)** and compute its derivative.

```
#1:   R(t) := [3·COS(t), SIN(t)]
#2:   R'(t)
#3:   [- 3·SIN(t), COS(t)]
```

We'll follow this with calculation of the unit tangent vector function and its derivative.

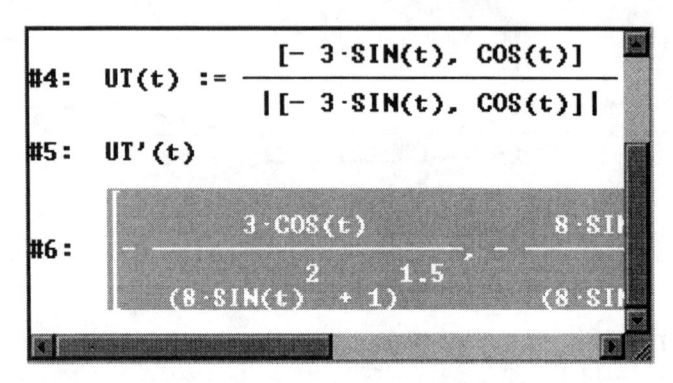

```
                  [- 3·SIN(t), COS(t)]
#4:   UT(t) := ─────────────────────────
                  |[- 3·SIN(t), COS(t)]|

#5:   UT'(t)

              3·COS(t)              8·SI
#6:   ─ ─────────────────────  , ─ ──────
              2       1.5
         (8·SIN(t)  + 1)            (8·SI
```

Now we compute the curvature by authoring the expression shown below.

$$\text{abs}(\#3)/\text{abs}(\#6)$$

Then clicking the ≈ icon yields

4.3. ARC LENGTH AND CURVATURE

#7: $\left\lVert \left[-\dfrac{3 \cdot \cos(t)}{\left((8 \cdot \sin(t))^2 + 1\right)^{1.5}}, -\dfrac{8 \cdot S}{(8 \cdot S} \right.\right.$

#8: $\sqrt{(64 \cdot \sin(t)^2 \cdot \cos(t)^4 + \cos(t)^2 \cdot (1}$

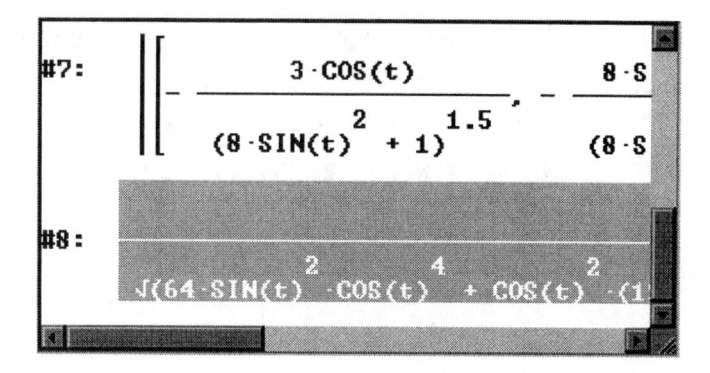

The plot below shows the graph of the curvature.

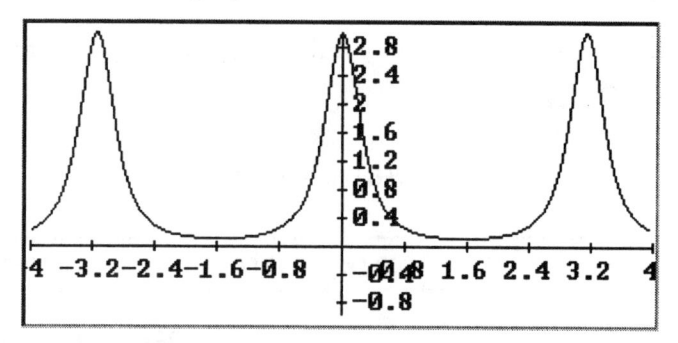

Notice that peaks in the curvature's graph occur at multiples of π. These correspond to points where the ellipse crosses the x-axis, which is where the ellipse curves the most–that is, changes direction most rapidly. The curvature is a minimum at each odd multiple of $\pi/2$. These correspond to points where the ellipse crosses the y-axis, which is where the ellipse curves the least.

Example 26 *Compute the curvature of the space curve described by*

$$\mathbf{r}(t) = \langle 2\cos(t), 2\sin(t), \sin(3t) \rangle.$$

Then plot the graph of the curvature for $0 \leq t \leq 2\pi$ and interpret its behavior in connection with the shape of the curve.

Let's first enter the parametrization as $\mathbf{r(t)}$ and compute its derivative.

#1: R(t) := [2·COS(t), 2·SIN(t), SIN(3·t)]

#2: R'(t)

#3: [-2·SIN(t), 2·COS(t), 3·COS(3·t)]

We'll follow this with calculation of the unit tangent vector and its derivative.

#4: $UT(t) := \dfrac{[-2 \cdot SIN(t), \; 2 \cdot COS(t), \; 3 \cdot COS(3 \cdot t)]}{|[-2 \cdot SIN(t), \; 2 \cdot COS(t), \; 3 \cdot COS(3 \cdot t)]|}$

#5: $UT'(t)$

#6: $\dfrac{2 \cdot \sqrt{2} \; (9 \cdot COS(t) \cdot COS(6 \cdot t) + 27 \cdot SIN(t) \cdot SIN(6 \cdot t) \cdots}{(9 \cdot COS(6 \cdot t) + 17)^{3/2}}$

Finally, we compute the curvature by authoring the following expression and clicking on the ≈ icon.

$$abs(\#6)/abs(\#3)$$

#7: $\left| \dfrac{2 \cdot \sqrt{2} \cdot (9 \cdot COS(t) \cdot COS(6 \cdot t) + 27 \cdot SIN(t) \cdots}{(9 \cdot COS(6 \cdot t) + 17\cdots} \right|$

#8: $\sqrt{2} \cdot \sqrt{(-81 \cdot COS(12 \cdot t) \cdot (81 \cdot COS(6 \cdot t) + 32) \cdots}$

A plot of the curvature is shown below.

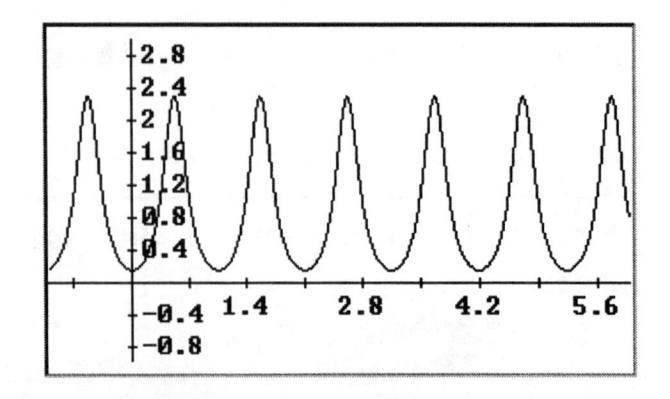

4.3. ARC LENGTH AND CURVATURE

Notice that the graph of the curvature has six peaks for t between 0 and 2π. We can see why these peaks occur if we **Load** the **Graphics.mth** utility file and use the **ISOMETRIC** and **axes** commands.

#9:   ISOMETRIC(R(t))

#10:   [2·SIN(t) - 2·COS(t), SIN(3·t) - COS(t) - SIN(t)]

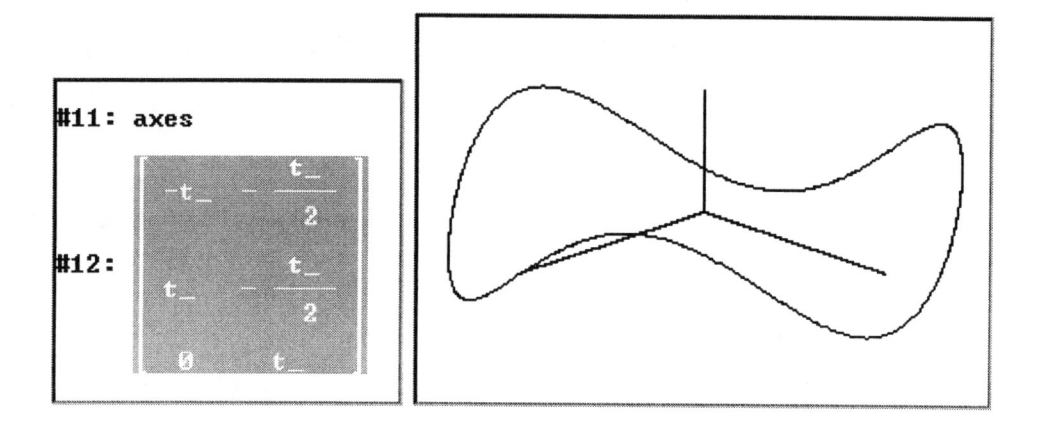

#11:   axes

#12: $\begin{bmatrix} -t\_ & -\dfrac{t\_}{2} \\ t\_ & \dfrac{t\_}{2} \\ 0 & t\_ \end{bmatrix}$

The space curve on the right above is the result of graphing #10 and #12 in a **2-D Plot** window with the **axes** turned off. Each peak in the curvature curve corresponds to a point on the space curve where the z-coordinate is a maximum or minimum, since these are (for this particular example) the points where the space curve changes direction most rapidly. The curvature is very close to zero at values of t corresponding to points where the space curve is nearly straight.

Example 27 *Determine, as a function of x, the curvature of a plane curve with equation $y = x^3$. Then apply the result to find and graph the curvature of $y = x^3$.*

First we need to parametrize the curve. Using x as the parameter, we have the simple parameterization

$$r(x) = \langle x, f(x) \rangle.$$

CHAPTER 4. VECTOR-VALUED FUNCTIONS

We enter this parametrization as **r(x)** and compute its derivative. Then we form the unit tangent vector.

#1: $R(x) := [x, x^3]$

#2: $R'(x)$

#3: $[1, 3 \cdot x^2]$

#4: $UT(x) := \dfrac{[1, 3 \cdot x^2]}{\|[1, 3 \cdot x^2]\|}$

#5: $UT'(x)$

#6: $\left[-\dfrac{18 \cdot x^3}{(9 \cdot x^4 + 1)^{3/2}}, \dfrac{6 \cdot x}{(9 \cdot x^4 + 1)^{3/2}} \right]$

Finally, we compute the curvature and simplify by clicking on the ≈ icon.

#7: $\dfrac{\left\| \left[-\dfrac{18 \cdot x^3}{(9 \cdot x^4 + 1)^{3/2}}, \dfrac{6 \cdot x}{(9 \cdot x^4 + 1)^{3/2}} \right] \right\|}{\|[1, 3 \cdot x^2]\|}$

#8: $\dfrac{6 \cdot |x|}{(9 \cdot x^4 + 1)^{3/2}}$

The plot below shows the graphs of $y = x^3$ and its curvature.

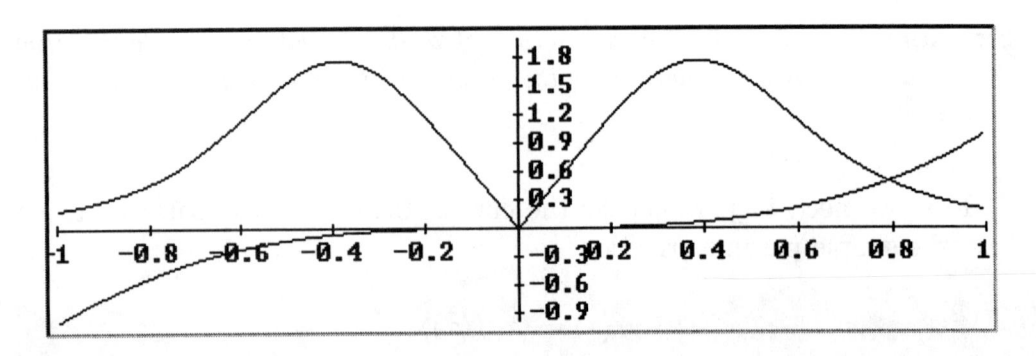

4.4 Velocity and Acceleration

- Section 10.4 in Stewart's **Multivariable Calculus: Concepts and Contexts** deals with velocity and acceleration.

Probably the most interesting and important application of parameterized curves is in the study of the motion of a moving object. We know from Stewart's text that if we interpret a vector valued function **r**(t) as the position vector of an object at time t, then the velocity of the object at time t is a vector quantity and is in fact the derivative of **r**(t)

$$\mathbf{v}(t) = \mathbf{r}'(t).$$

Also, the length of the velocity vector is the speed of the object, and the derivative of velocity is acceleration, again a vector quantity:

$$\mathbf{a}(t) = \mathbf{v}'(t) = \mathbf{r}''(t).$$

Example 28 *Suppose that the position vector of an object moving in the plane is*

$$\mathbf{r}(t) = \langle 2\cos(.17t) + \cos(.83t),\ 2\sin(.17t) + \sin(.83t) \rangle.$$

Compute the velocity and acceleration as functions of t. Then plot the curve for $0 \le t \le 4\pi$ along with velocity and acceleration vectors on that interval at multiples of $\pi/3$. Comment on the relationship between the change in the direction of the object's path and the direction of the acceleration vector.

The first step is to enter the parametrization as **r(t)** and then use the **VECTOR** command to create the matrix necessary to plot the velocity vectors.

```
#1:  R(t) := [2·COS(0.17·t) + COS(0.83·t), 2·SIN(0.17·t)

#2:  VECTOR([R(n·π/3), R(n·π/3) + R'(n·π/3)], n, 0, 5, 1)

#3:  [[ 3       0    ]   [ 2.61384  1.11796 ]   [ 1.70779  1.68
     [ 3       1.17  ] , [ 1.91968  1.98832 ] , [ 0.770903 1.86
```

Then we use the **VECTOR** command to create the matrix necessary to plot the acceleration vectors.

$$\#4: \quad \text{VECTOR}\left(\left[R\left(\frac{n\cdot\pi}{3}\right), R\left(\frac{n\cdot\pi}{3}\right) + R''\left(\frac{n\cdot\pi}{3}\right)\right], n, 0, 5, 1\right)$$

$$\#5: \quad \left[\begin{matrix} 3 & 0 \\ 2.25330 & 0 \end{matrix}\right], \left[\begin{matrix} 2.61384 & 1.11796 \\ 2.11230 & 0.581550 \end{matrix}\right], \left[\begin{matrix} 1.70779 & 1. \\ 1.76850 & 0. \end{matrix}\right]$$

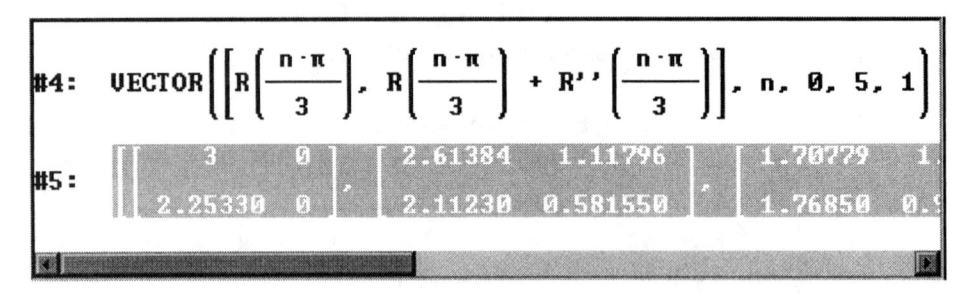

Finally, we plot #1, #3 and #5 in a **2-D Plotting** window.

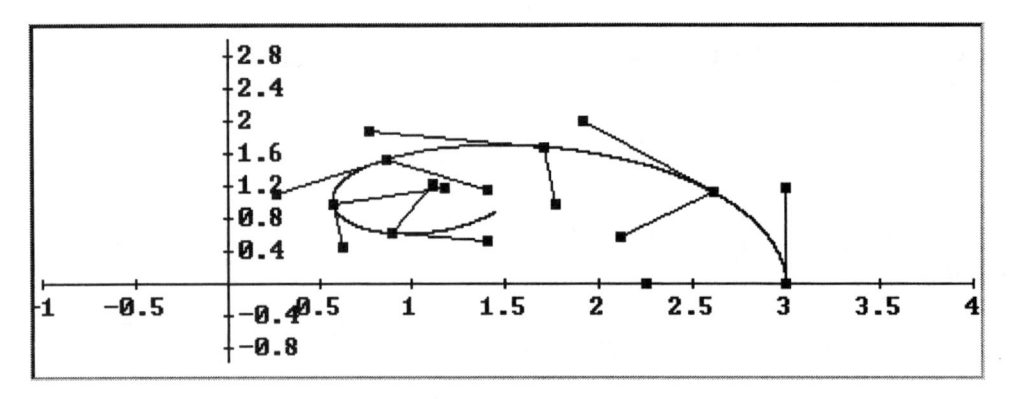

Comment: *The path of the object always bends in the direction of the acceleration vector. In this example, that's always a bend to the left. The acceleration vectors appear sometimes to be orthogonal to the path, but* **not** *always.*

Exercises:

1. For each of the curves described in Exercises 1-8 of Section 4.1, plot the curve and compute its arc length (exactly if possible).

2. For the curves described in Exercises 3 and 8 of Section 4.1, plot the parametrization's arc length function.

3. For each of the curves described in Exercises 1-8 of Section 4.1, find the curvature and plot its graph.

4. Find the point(s) at which the graph of $y = x^4$ has maximum curvature.

5. Find the equation of the circle that best approximates the parabola $y = x^2$ at $(1/2, 1/4)$ in the sense that it is tangent to and has the same curvature as the parabola at that point. Create a plot of both the parabola and the circle. (Note: The curvature of a circle with radius R is $1/R$.)

6. For each of the curves described in Exercises 1–8 of Section 4.1, plot the curve along with a sampling of unit tangent vectors.

7. For each of the curves described in Exercises 1–8 of Section 4.1, plot the curve along with a sampling of acceleration vectors.

8. There are two definitions that one might give for a scalar acceleration of an object moving according to a parametrization $\mathbf{r}(t)$. One is $|\mathbf{r}''(t)|$, the length of the acceleration vector. The other is the derivative of the speed. Give an example that shows that these definitions, in general, produce different results.

4.5 Parametric Surfaces

- Parametric surfaces are the subject of Section 10.5 of Stewart's **Multivariable Calculus: Concepts and Contexts**.

We have seen that space curves have parameterizations that take the form of vector valued functions of one variable. Analogous parameterizations of *surfaces* take the form of vector valued functions of *two* variables (i.e., two parameters):
$$\mathbf{r}(s,t) = \langle x(s,t), y(s,t), z(s,t) \rangle.$$

The **Grahpics.mth** file utility contains the commands **ISOMETRICS** and **COPROJECTION** which can be used to plot a parameterized surface in a **2-D Plotting** window.

Example 29 *The usual parametrization of the unit sphere is*
$$\mathbf{r}(\theta, \phi) = \langle \cos(\theta) \sin(\phi), \sin(\theta) \sin(\phi), \cos(\phi) \rangle$$
for $0 \leq \theta \leq 2\pi$ and $0 \leq \phi \leq \pi$. The parameters θ and ϕ are the familiar spherical coordinate angles. Note that the length of $\mathbf{r}(\theta, \phi)$ is always 1, since
$$\cos^2(\theta) \sin^2(\phi) + \sin^2(\theta) \sin^2(\phi) + \cos^2(\phi) = 1.$$

To plot the sphere, we **Load** the utility file **Graphics.mth**. Then we define the function **r** describing the parameterization and apply the **ISO-METRICS** command.

```
#1:   R(s, t) := [COS(s)·SIN(t), SIN(s)·SIN(t), COS(t)]
#2:   ISOMETRICS(R(s, t), s, 0, 6.3, 10, t, 0, 3.14, 10)
```

followed by clicking on the ≈ icon. We have omitted the output (due to its size). Then we apply the **COPROJECTION** command to the output in #3 and again click on the ≈ icon.

COPROJECTION(#3)

Again, the output has been omitted. Then, we create a **2-D Plot** window, turn off the **axes**, and set the **Points** (from within the **Options** menu) as shown below.

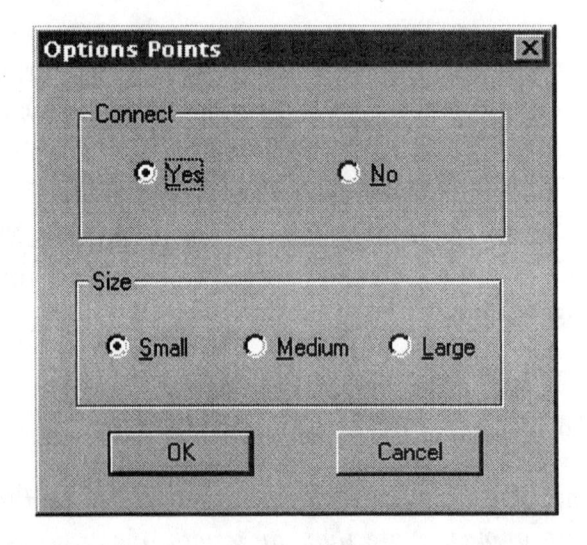

Finally, we plot #3 and #5. In the **2-D Plot** window.

4.5. PARAMETRIC SURFACES

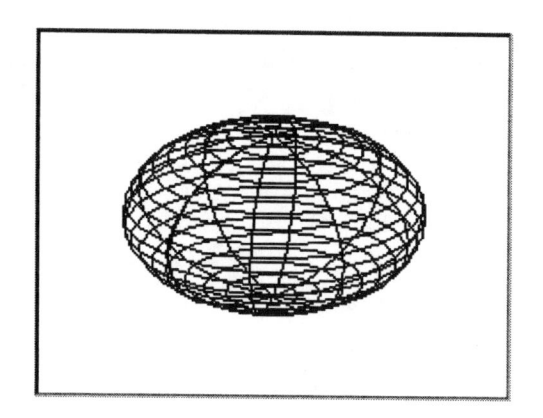

Exercises:

1. Given a plane curve parametrized by

$$\mathbf{q}(t) = \langle f(t), g(t) \rangle,$$

a parametrization of the corresponding surface of revolution about the y-axis is given by

$$\mathbf{r}(\theta, t) = \langle f(t)\cos\theta, \ f(t)\sin\theta, \ g(t) \rangle.$$

Plot the plane curve and then **Load** the **Graphics.mth** utility and use the **ISOMETRICS** and **COPROJECTION** commands to create a wire-frame surface of revolution corresponding to:

$$\mathbf{q}(t) = \langle 2 + \cos(t), \ \sin(t) \rangle, \quad 0 \le t \le 2\pi.$$

2. Repeat problem 1 for

$$\mathbf{q}(t) = \langle t, \ t^2 \rangle, \quad 0 \le t \le 1.$$

3. Repeat problem 1 for

$$\mathbf{q}(t) = \langle 2\sin(t) + \sin(5t), \ 2\cos(t) + \cos(5t) \rangle, \quad -\pi/2 \le t \le \pi/2.$$

Chapter 5

Multivariate Functions

This chapter discusses the use of **DERIVE** to analyze real valued functions of two or more variables. **DERIVE** has the power and flexibility to explore limits, continuity, differentiation and optimization for functions of several variables.

We have already encountered such functions in our brief investigation of surfaces in Chapter 3. You should review Chapter 3 before proceeding further.

- Section 11.1 of Stewart's **Multivariable Calculus: Concepts and Contexts** discusses multivariate functions and related notions, including surfaces and contours.

5.1 Limits and Continuity

- The notions of limit and continuity for multivariate functions are developed in Section 11.2 of Stewart's **Multivariable Calculus: Concepts and Contexts**. Be sure to review these notions before working in this section.

The following examples illustrate how **DERIVE** can be used to explore some of the strange behaviors which can occur with functions of two or more variables.

Example 30 *Consider the function*
$$f(x,y) = (y-x)/\sqrt{x^2+y^2}$$

which is defined and continuous at all points in the xy-plane except $(0,0)$, where it is undefined. Explore the behavior of this function near $(0,0)$.

It will be convenient to define $f(x, y)$ as a function in **DERIVE**.

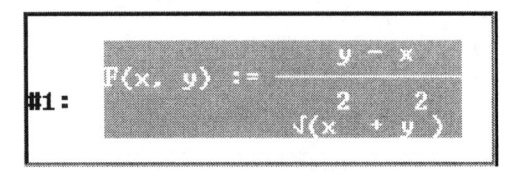

#1: $\quad F(x, y) := \dfrac{y - x}{\sqrt{x^2 + y^2}}$

If we create a graph in a **3-D Plot** window for x and y between -1 and 1 then we see

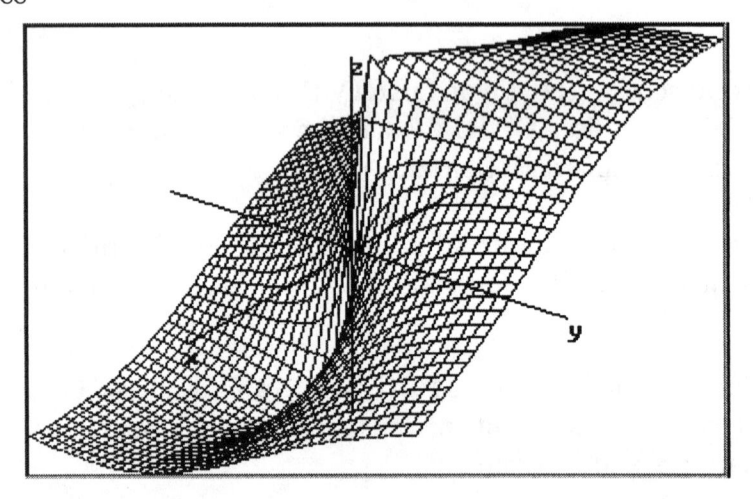

Obviously something strange is happening near $(x, y) = (0, 0)$. We can investigate the behavior of f near $(0,0)$ in a number of ways. For example, we might consider the approach to $(0,0)$ along the x-axis. To see this, we need to plot $f(x, 0)$. We create the expression $f(x, 0)$ below and show its plot in a **2-D Plot Window**.

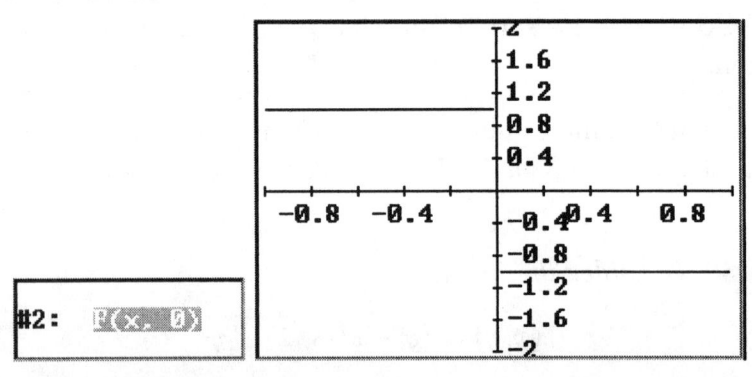

#2: $\quad F(x, 0)$

5.1. LIMITS AND CONTINUITY

This indicates that $f(x, y)$ does not have a limit as (x, y) approaches $(0, 0)$ along the x-axis. We can confirm this by using the **DERIVE** command **LIM**.

$$\#3: \quad \lim_{x \to 0} F(x, 0) = \pm 1$$

Consequently, we can already see that

$$\lim_{(x,y) \to (0,0)} f(x, y) \text{ does not exist.}$$

However, it turns out that this only tells a portion of the story. Let's see what happens when we force (x, y) to approach $(0, 0)$ along other lines through the origin. For example, look at the result of restricting $f(x, y)$ to the line $y = mx$.

$$\#4: \quad F(x, m \cdot x) = \frac{(m - 1) \cdot \text{SIGN}(x)}{\sqrt{(m^2 + 1)}}$$

Example 31 *Now let's consider the function*

$$f(x, y) = \frac{-x}{x^2 + y^2}$$

We can see that this function is defined and continuous at all points in the xy-plane except $(0, 0)$. Let's explore the behavior of $f(x, y)$ for (x, y) near $(0, 0)$.

Once again, it is convenient to define the function $f(x, y)$.

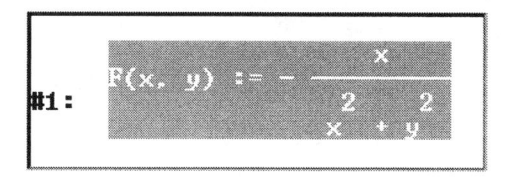

$$\#1: \quad F(x, y) := -\frac{x}{x^2 + y^2}$$

Then we can obtain a graph of $f(x, y)$ in a **3-D Plot** window.

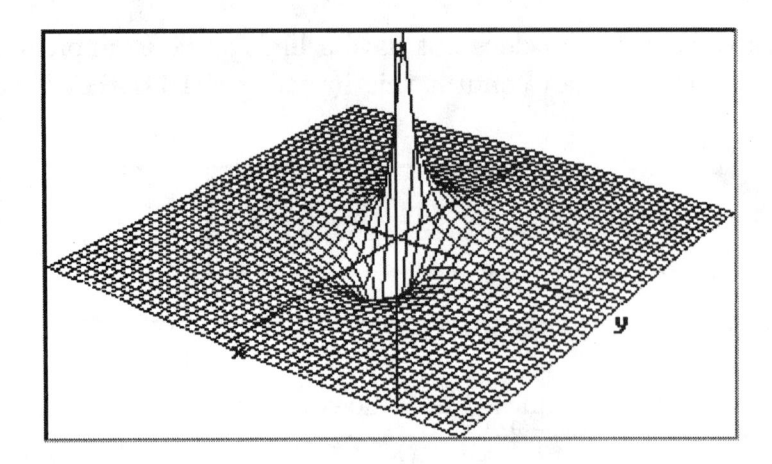

We can see that the behavior near $(0,0)$ is a little strange. Note that $f(x,0) = -1/x$ and $f(x, mx) = -1/\left((1+m^2)x\right)$.

#2: $F(x, 0) = -\dfrac{1}{x}$

#3: $F(x, m \cdot x) = -\dfrac{1}{x \cdot (m^2 + 1)}$

Consequently, $f(x,y)$ becomes unbounded when (x,y) approaches $(0,0)$ along lines of the form $y = mx$. However, this does not imply that $f(x,y)$ becomes unbounded when (x,y) approaches $(0,0)$ (along arbitrary directions). This can be seen by considering approaches along $x = -y^2$ and $x = -y^3$.

#4: $F(-y^2, y) = \dfrac{1}{y^2 + 1}$

#5: $F(-y^3, y) = \dfrac{y}{y^4 + 1}$

These functions are graphed below (note that the horizontal axis represents y).

5.1. LIMITS AND CONTINUITY

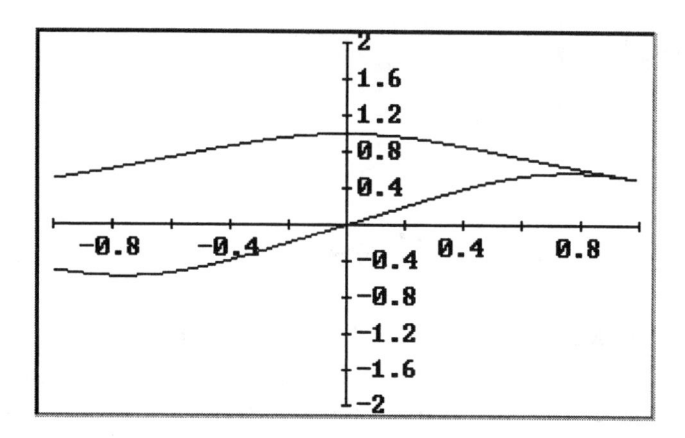

Notice that the approach to $(0,0)$ along $x = -y^2$ yields a value of 1 and the approach to $(0,0)$ along $x = -y^3$ yields 0. This function clearly has no limit as (x, y) approaches $(0, 0)$. In fact, this function behaves very strangely near $(0, 0)$.

Exercises:

1. Consider the behavior of the function
$$f(x,y) = \frac{xy}{x^2 + y^2}$$
for (x, y) near $(0, 0)$ by evaluating $f(x, y)$ along lines through the origin. You should find that $f(x, y)$ is a constant function along lines through the origin. What will this imply about the contour plot of $f(x, y)$ near $(0, 0)$? Explain. Generate a contour plot to confirm your answer.

2. Create contour plots for each of the functions in Examples 1 and 2 of this section. Notice that the contours seem to intersect at the origin in each case. Are they actually intersecting? Explain why distinct contours can only appear to intersect at a point where a function is discontinuous.

3. Investigate the behavior of each of the following functions near $(0, 0)$. Your analysis should include restrictions of the functions to different curves through the origin and plots of the behavior of the function along these curves. A contour plot might help you find appropriate curves to work with.

   a) $f(x,y) = \frac{x^2 - y^2}{x^4 + y^4}$

b) $f(x,y) = \frac{x^2-y^2}{x^2+y^2}$

c) $f(x,y) = \frac{\sin(x^2+y^2)}{x^2+y^2}$

5.2 Partial Derivatives

- A detailed discussion of partial derivatives of multivariate functions can be found in Section 11.3 of Stewart's **Multivariable Calculus: Concepts and Contexts**. Read Stewart's Section 11.3 carefully before considering the examples in this section.

The examples in this section demonstrate how **DERIVE** can be used to work with partial derivatives of a function of two or more variables.

Example 32 *Let's graph the function*

$$f(x,y) = 4e^{-(x^2+y^2)} + \frac{y^2}{6}$$

and give an interpretation of the partial derivatives

$$\frac{\partial}{\partial x}f(1/4, 1/4) \text{ and } \frac{\partial}{\partial y}f(1/4, 1/4).$$

We start by defining $f(x,y)$.

```
#1:    F(x, y) := 4·EXP(- (x^2 + y^2)) + y^2/6
```

Since our interest lies near $(1/4, 1/4)$, we graph $z = f(x,y)$ for x and y near $(1/4, 1/4)$.

5.2. PARTIAL DERIVATIVES

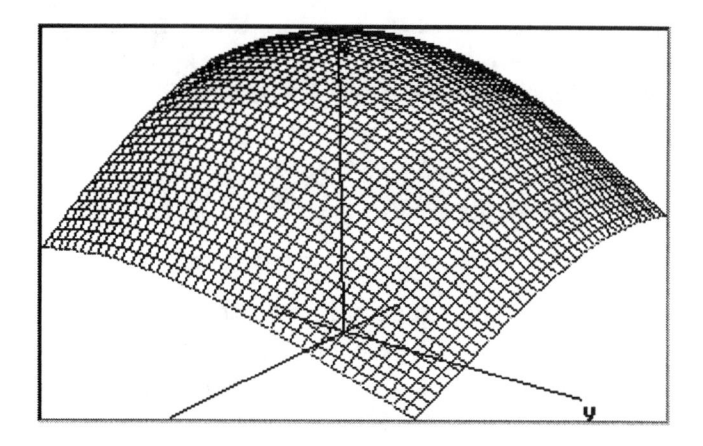

Notice the lines in the wire frame which appear to run parallel to the x and y axes. These are lines along which x and y respectively are constant. These lines help use interpret the partial derivatives $\frac{\partial}{\partial x}f(1/4, 1/4)$ and $\frac{\partial}{\partial y}f(1/4, 1/4)$. This is because the partial derivative $\frac{\partial}{\partial x}f(1/4, 1/4)$ represents the change in $f(x, 1/4)$ with respect to x at $x = 1/4$. Similarly, the partial derivative $\frac{\partial}{\partial y}f(1/4, 1/4)$ represents the change in $f(1/4, y)$ with respect to y at $y = 1/4$. Each of these values is computed below by first using the **DIF** command to create the partial derivatives, and then defining functions **DXF** and **DYF** which represent the partial derivatives.

#2: $\quad \dfrac{d}{dx} F(x, y) = -8 \cdot x \cdot \hat{e}^{-x^2 - y^2}$

#3: $\quad \text{DXF}(x, y) := -8 \cdot x \cdot \hat{e}^{-x^2 - y^2}$

#4: $\quad \dfrac{d}{dy} F(x, y) = \dfrac{y}{3} - 8 \cdot y \cdot \hat{e}^{-x^2 - y^2}$

#5: $\quad \text{DYF}(x, y) := \dfrac{y}{3} - 8 \cdot y \cdot \hat{e}^{-x^2 - y^2}$

The values of $\frac{\partial}{\partial x}f(1/4, 1/4)$ and $\frac{\partial}{\partial y}f(1/4, 1/4)$ are shown below.

```
                  ⎛ 1   1 ⎞         - 1/8
#6:    DXF ⎜ ─ , ─ ⎟  = - 2·ê
                  ⎝ 4   4 ⎠

                  ⎛ 1   1 ⎞     1           - 1/8
#7:    DYF ⎜ ─ , ─ ⎟  = ── - 2·ê
                  ⎝ 4   4 ⎠    12
```

Consequently, the slope of the tangent line to the graph of $z = f(x, 1/4)$ at $x = 1/4$ is approximately -1.77, and the slope of the tangent line to the graph of $z = f(1/4, y)$ at $y = 1/4$ is approximately -1.67.

DERIVE can also be used to compute higher order partial derivatives.

Example 33 *Consider the function given by*

$$g(x, y, z) = 2xy - 3y^2 z + 2z^2$$

The screen shot below shows the definition of $g(x, y, z)$ along with the computation of the partial derivatives

$$\frac{\partial}{\partial x} g(x, y, z) \text{ and } \frac{\partial^3}{\partial z \partial y^2} g(x, y, z).$$

```
                                                2            2
#1:    G(x, y, z)  :=  2·x·y  -  3·y ·z  +  2·z

        d
#2:    ── G(x, y, z) = 2·y
       dx

        d  ⎛ d ⎞2
#3:    ── ⎜── ⎟   G(x, y, z) = -6
       dz ⎝dy ⎠
```

The last line is the result of authoring the expression

$$\text{DIF}(\text{DIF}(\text{g}(\text{x}, \text{y}, \text{z}), \text{y}, 2), \text{z}) =$$

Exercises:

1. Let $f(x, y) = x^2 + y^2$. Graph this function for $-1 \leq x \leq 1$ and $-1 \leq y \leq 1$. Then interpret the partial derivatives $\frac{\partial}{\partial x} f(.5, .5)$ and $\frac{\partial}{\partial y} f(.5, .5)$.

5.3. THE TANGENT PLANE AND LINEAR APPROXIMATION

2. A function is said to be harmonic provided it satisfies *Laplace's* equation:
$$u_{xx} + u_{yy} = 0.$$
Show that the following functions are harmonic.

    a) $u(x,y) = e^{-x} \sin(y)$
    b) $u(x,y) = \tan^{-1}(y/x)$
    c) $u(x,y) = \sqrt{x^2 + y^2}.$

3. Verify that
$$u(t,x,y) = e^{-2t} \sin(x)\cos(y) - 3e^{-4t}\sin(\sqrt{2}x)\cos(\sqrt{2}y)$$
satisfies the heat equation
$$u_t - (u_{xx} + u_{yy}) = 0.$$

5.3 The Tangent Plane and Linear Approximation

- The tangent plane and linear approximation are discussed in Section 11.4 of Stewart's **Multivariable Calculus: Concepts and Contexts**. The basic concepts include the equation for a tangent plane to the graph of $z = f(x,y)$ at the point $(a, b, f(a,b))$, and the linear approximation (or linearization) of $f(x,y)$ at (a,b). The tangent plane is given by
$$z = f(a,b) + f_x(a,b)(x-a) + f_y(a,b)(y-b)$$
and the linear approximation is given by
$$\Lambda_{(a,b)}(x,y) = f(a,b) + f_x(a,b)(x-a) + f_y(a,b)(y-b)$$
Note that the tangent plane is simply the graph of the linear approximation.

In this section we use **DERIVE** to discuss tangent planes and linear approximation.

Example 34 *Find the equation for the tangent plane to the graph of* $z = 3 - (x^2 + y^2)$ *at the point* $\left(\frac{1}{2}, \frac{1}{3}, \frac{95}{36}\right)$.

We start by defining the function $f(x, y) = 3 - (x^2 + y^2)$ and defining the partial derivatives as functions.

```
#1:   F(x, y) := 3 - (x² + y²)
#2:   GRAD(F(x, y), [x, y]) = [- 2·x, - 2·y]
#3:   DXF(x, y) := - 2·x
#4:   DYF(x, y) := - 2·y
```

Finally, we verify that $f\left(\frac{1}{2}, \frac{1}{3}\right) = \frac{95}{36}$

```
#5:   F(1/2, 1/3) = 95/36
```

and we compute the equation of the tangent plane by authoring the expression

$$z = f(1/2, 1/3) + \mathbf{DXF}(1/2, 1/3)(x - 1/2) + \mathbf{DYF}(1/2, 1/3)(y - 1/3)$$

and clicking the ▨ icon.

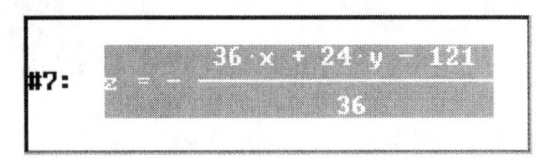

```
#7:   z = - (36·x + 24·y - 121)/36
```

Now, since the tangent plane to the graph is simply the linear approximation, we can use the tangent plane to approximate $f(0.4, 0.3)$ and compare it with the function value given by **DERIVE**. We start this process by defining a function $TP(x, y)$ which gives the tangent plane approximation (or linear approximation) at $(1/2, 1/3)$. Then we evaluate each of f and TP at $(.4, .3)$ and click on the ▨ icon.

5.3. THE TANGENT PLANE AND LINEAR APPROXIMATION   81

#8: $TP(x, y) := -\dfrac{36 \cdot x + 24 \cdot y - 121}{36}$

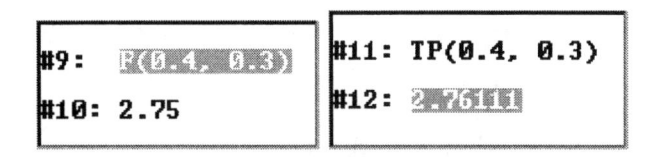

#9:   (0.4, 0.3)

#10: 2.75

#11: TP(0.4, 0.3)

#12: 2.76111

We can see from the given values that the linear approximation does a very good job of approximating the function value. This will be true for every point (x, y) close to $(1/2, 1/3)$.

Exercises:

1. Find the equation of the tangent plane to the graph of $z = f(x, y)$ at the point $(a, b, f(a, b))$ for each of the choices of $f(x, y)$ and (a, b) given below.

   (a) $f(x, y) = xy$ at the point $(-1, -1, f(-1, -1))$
   (b) $f(x, y) = 2x \sin(y) + x \cos(y) - x^2/100$ at the point $(3, 0, f(3, 0))$.

2. Find the linearization of
$$f(x, y) = x^{-y^{-x}} + y^{3x-y}$$
at $(2, 1)$ and use it to approximate $f(1.94, 1.04)$. Compute the relative error of your approximation.

3. Find the linear approximation to the function
$$f(x, y, z) = \sin(x \cos(y \sin(z)))$$
at $\left(\dfrac{\pi}{3}, \dfrac{2\pi}{3}, \dfrac{5\pi}{6}\right)$. Use this to approximate $f(1, 2, 5)$ and compute the relative error in the approximation.

5.4 Directional Derivatives and the Gradient

- Section 11.6 in Stewart's **Multivariable Calculus: Concepts and Contexts** contains formal definitions and thorough discussion of the directional derivative and the gradient vector.

Recall that for a function $f(x,y)$ the gradient is given by
$$\nabla f(x,y) = \langle f_x(x,y), f_y(x,y) \rangle$$
and for a function $f(x,y,z)$ the gradient is given by
$$\nabla f(x,y,z) = \langle f_x(x,y,z), f_y(x,y,z), f_z(x,y,z) \rangle.$$
The directional derivative of $f(x,y)$ at (a,b) in the direction of a nonzero vector $\mathbf{v}$ is given by
$$D_u f(a,b) = \nabla f(a,b) \cdot \frac{\mathbf{v}}{|\mathbf{v}|}$$
and the directional derivative of $f(x,y,z)$ at (a,b,c) in the direction of a nonzero vector $\mathbf{v}$ is given by
$$D_u f(a,b,c) = \nabla f(a,b,c) \cdot \frac{\mathbf{v}}{|\mathbf{v}|}$$

Finally, if $\nabla f(P) \neq 0$ then $\nabla f(P)$ points in the direction of greatest increase in f from the point P, and $-\nabla f(P)$ points in the direction of greatest decrease in f from the point P. In this case, the largest directional derivative at P is given by $|\nabla f(P)|$ and the least directional derivative at P is given by $-|\nabla f(P)|$. Moreover, $\nabla f(P)$ is perpendicular to the contour of f that passes through P at the point P.

Example 35 *Let $f(x,y) = x\cos(xy) + 3(1 - \sin(x+y))$. Let's find the directional derivative at $(\pi/3, \pi/4)$ in the direction of the vector $\langle 2, 1 \rangle$.*

We start by defining $f(x,y)$ and creating the gradient vector.

```
#1:   F(x, y) := x·COS(x·y) + 3·(1 - SIN(x + y))

#2:   GRAD(F(x, y), [x, y]) = [COS(x·y) - x·y·SIN

#3:   GRADF(x, y) := [COS(x·y) - x·y·SIN(x·y) - 3
```

5.4. DIRECTIONAL DERIVATIVES AND THE GRADIENT

Then we convert $\langle 2, 1 \rangle$ to a unit vector and form the directional derivative.

```
#4:    u :=  [2, 1]
            ───────
            | [2, 1] |

#5:    GRADF ( π/3, π/4 ) · u

#6:    0.7518172
```

Note that the entry in #6 was created by clicking on the [≈] icon.

Exercises:

1. Find the directional derivative of each of the following functions $f(x, y)$ at the specified point (a, b) and in the direction of

$$\mathbf{u} = \langle \cos(\theta), \sin(\theta) \rangle$$

Plot the resulting function of θ for $0 \leq \theta \leq 2\pi$, and find where the extreme values occur. Then, for those values of θ, verify that $\mathbf{u}$ corresponds to either $\pm \nabla f(a, b)$.

  (a) $f(x, y) = \frac{1}{\sqrt{x^2+y^2}}$ at the point $(1, 1)$.

  (b) $f(x, y) = x^2 y^3$ at the point $(1, 1)$.

  (c) $f(x, y) = \cos(xy)$ at the point $(1/4, \pi)$.

2. Compute the directional derivative of $f(x, y, z)$ at the given point (a, b, c) in the direction of the given vector.

  (a) $f(x, y, z) = \frac{x-3yz}{\sqrt{x^2+y^2+z^2}}$ at the point $(1, 1, -2)$ in the direction of the vector $\langle 2, 2, 3 \rangle$.

  (b) $f(x, y, z) = x^2 y^3 z - 2xy^2 z^3$ at the point $(-1, 1, 1)$ in the direction of the vector $\langle 1, 1, -2 \rangle$.

3. Find the direction of greatest increase of f at the given point. Then find the greatest directional derivative for f at the given point.

(a) $f(x,y) = \frac{1}{\sqrt{x^2+y^2}}$ at the point $(1,1)$.

(b) $f(x,y) = x^2 y^3$ at the point $(1,1)$.

(c) $f(x,y) = \cos(xy)$ at the point $(1/4, \pi)$.

(d) $f(x,y,z) = \frac{x-3yz}{\sqrt{x^2+y^2+z^2}}$ at the point $(1,1,-2)$.

(e) $f(x,y,z) = x^2 y^3 z - 2xy^2 z^3$ at the point $(-1,1,1)$.

4. Give a simple geometric argument for why the gradient of $f(x,y) = x^2 + y^2$ at (a,b) is orthogonal to the contour of f through (a,b). Include an illustration consisting of a plot of the unit circle and several gradient vectors with initial points at the origin.

5. Define the function $f(x,y) = y - \frac{1}{2}x^2$ and plot the contour $f(x,y) = 0$ for $-3 \le x \le 3$ and $0 \le y \le 3$ by graphing $y = \frac{1}{2}x^2$. Then determine five points (a_i, b_i) for $i = 1, 2, 3, 4, 5$ which lie on this contour. Finally, plot $\nabla f(a_i, b_i)$ with initial end at (a_i, b_i) for $i = 1, 2, 3, 4, 5$ on top of your graph. The resulting plot should verify that $\nabla f(a_i, b_i)$ is perpendicular to the contour at the point (a_i, b_i).

5.5 Optimization and Lagrange Multipliers

- Optimization is discussed in Section 11.7 of Stewart's **Multivariable Calculus: Concepts and Contexts**. The method of Lagrange multipliers is discussed in Section 11.8. In particular, critical points and classification of critical points is discussed in Section 11.7. In addition, optimization of differentiable functions on closed bounded regions is also discussed in Section 11.7. Optimization of differentiable functions subject to constraints is discussed in Section 11.8.

Example 36 *Let's find and classify all of the critical points of*

$$f(x,y) = x^4 - 3xy + 2y^2$$

We start by defining the function $f(x,y)$ and computing the partial derivatives of f.

5.5. OPTIMIZATION AND LAGRANGE MULTIPLIERS

```
#1:  F(x, y) := x^4 - 3·x·y + 2·y^2

#2:  GRAD(F(x, y), [x, y]) = [4·x^3 - 3·y, 4·y - 3·x]
```

Then we set each of the partials equal to zero and plot the results in a **2-D Plot** window.

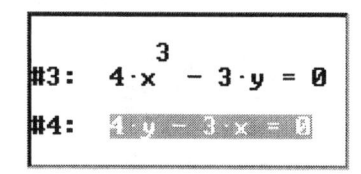

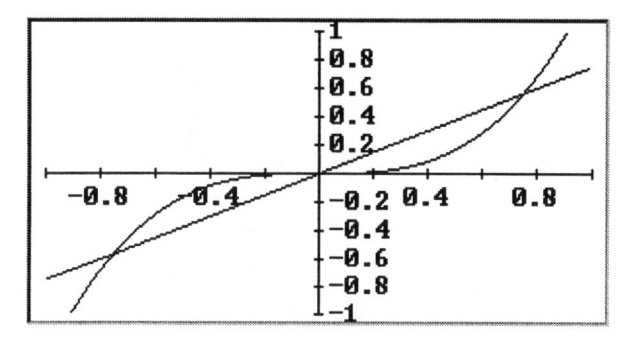

```
#3:  4·x^3 - 3·y = 0
#4:  4·y - 3·x = 0
```

To find the critical points of $f(x, y)$ we need to solve

$$f_x(x, y) = 0$$
$$f_y(x, y) = 0$$

We can see from the equations above and the graph that $(x, y) = (0, 0)$ is a solution of this system. Also, the graph shows that there are exactly three solutions of this system, and from symmetry, we see that if (x, y) solves the system then so does $(-x, -y)$. There are a couple of ways to find the other solutions. For this simple system, we could simply solve the first equation for y in terms of x and substitute the result into the second equation. Then we could solve the second equation for x. The method that we use below is more general, and does not rely upon the system having this simple algebraic structure. We start by **Loading** the **Solve.mth** utility file. Then we use the **NEWTONS** command with an initial guess of $(x, y) = (1, 1)$ (since this point is close to the true solution) and request 6 iterations, followed by clicking on the ≈ icon.

#5: `NEWTONS([4·x^3 - 3·y, 4·y - 3·x], [x, y], [1, 1], 6)`

Then we extract the last row of the output and write the fraction equivalent.

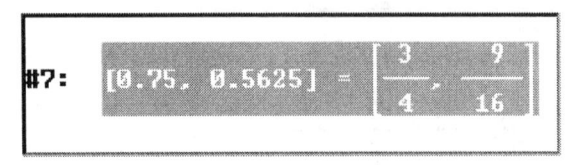

#7: $[0.75, 0.5625] = \left[\dfrac{3}{4}, \dfrac{9}{16}\right]$

The output above corresponds to the last iteration of the **NEWTONS** command. Since the last two rows from **NEWTONS** command give the same values, this must correspond to a solution of the system. Therefore, the critical points are $(3/4, 9/16)$, $(0,0)$ and $(-3/4, -9/16)$. We can classify these critical points by checking the second derivative test. However, let's first look at a contour plot of $f(x,y)$. Author the expression shown below and click on the ▪ icon to create a list of equations used to create the contour plot in a **2-D Plot** window.

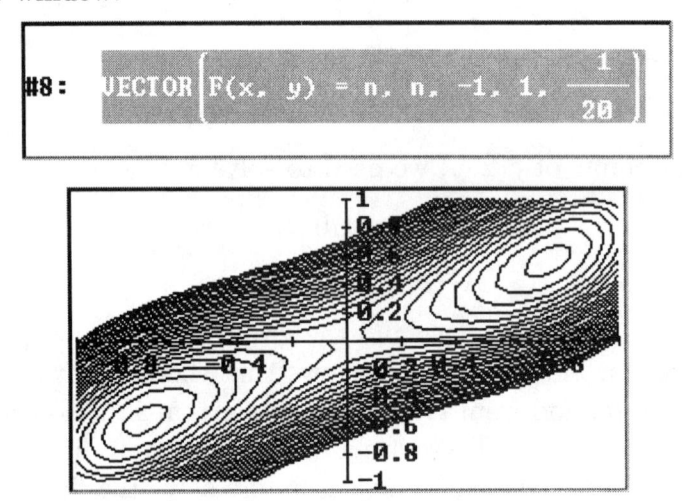

#8: `VECTOR(F(x, y) = n, n, -1, 1, 1/20)`

We can see from the graph that $(0,0)$ must be a saddle point, and the points $(3/4, 9/16)$ and $(-3/4, -9/16)$ must be either local maximums or local minimums. We can determine the exact nature of these critical points by creating the function

$$TEST(x,y) = \left[\dfrac{\partial^2 f(x,y)}{\partial x^2}, \dfrac{\partial^2 f(x,y)}{\partial x^2}\dfrac{\partial^2 f(x,y)}{\partial y^2} - \left(\dfrac{\partial^2 f(x,y)}{\partial x \partial y}\right)^2\right].$$

5.5. OPTIMIZATION AND LAGRANGE MULTIPLIERS

This can be done in **DERIVE** by using the **DIF** command. The result is shown below followed by its evaluation at the three critical points given above..

```
#17:  [12·x², 48·x² - 9]

#18:  TEST(x, y) := [12·x², 48·x² - 9]
```

```
#19:  TEST(0, 0) = [0, -9]

#20:  TEST(3/4, 9/16) = [27/4, 18]

#21:  TEST(-3/4, -9/16) = [27/4, 18]
```

Apparently, $(0,0)$ is a saddle point (as stated above), and each of $(3/4, 9/16)$ and $(-3/4, -9/16)$ are local minimums.

Example 37 *Consider the function*

$$f(x,y) = \left(y^2 - x + xy + \ln\left(x^2 + y^4 + 1\right)\right) e^{-(x^2+y^2)}$$

for $-2 \leq x \leq 2$ *and* $-2 \leq y \leq 2$. *Find all of the critical points of* f.

We start by defining the function $f(x, y)$ and using the **VECTOR** command to form a matrix which can be used to create a contour plot of $f(x, y)$.

```
#1:  F(x, y) := (y² - x + x·y + LN(x² + y⁴ +

#2:  F(0, 0) = 0

#3:  VECTOR[[F(x, y) = n], n, -1, 1, 1/10]
```

Clicking on the ▇ icon, followed by creating a **2-D Plot** window, gives the plot below.

CHAPTER 5. MULTIVARIATE FUNCTIONS

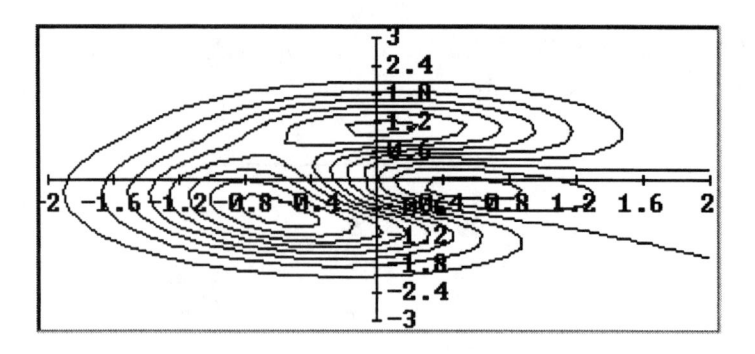

From the contour plot, we can see that f has four critical points in the region. Three of these are either local maximums or local minimums, and one of these is a saddle point. Based upon the contours, we might guess that the critical points occur near $(-.5, -.75)$, $(0, 1)$, $(.5, -.2)$ and $(-.5, .5)$. To find the actual values we set the first order partials equal to zero. We start by using the **DIF** command to form the expression in #4 below, and then clicking on the [=] icon.

#4: $\left[\dfrac{d}{dx} F(x, y), \dfrac{d}{dy} F(x, y)\right]$

#5: $\left[-\hat{e}^{-x^2 - y^2} \cdot \left(2 \cdot x \cdot \text{LN}(x^2 + \ldots\right.\right.$

Then we **Load** the **Solve.mth** utility file and use the **NEWTONS** command with the initial guessed given above. For example, if we **Author** the **Expression**

$$\text{NEWTONS}(\#5, [x,y], [-.5, -.75], 6)$$

followed by clicking on the [≈] icon, then we obtain the matrix below.

5.5. OPTIMIZATION AND LAGRANGE MULTIPLIERS

#7: $\begin{bmatrix} -0.5 & -0.75 \\ -0.610135 & -0.711347 \\ -0.597085 & -0.729382 \\ -0.597088 & -0.729060 \\ -0.597067 & -0.729096 \\ -0.597065 & -0.729098 \\ -0.597068 & -0.729095 \end{bmatrix}$

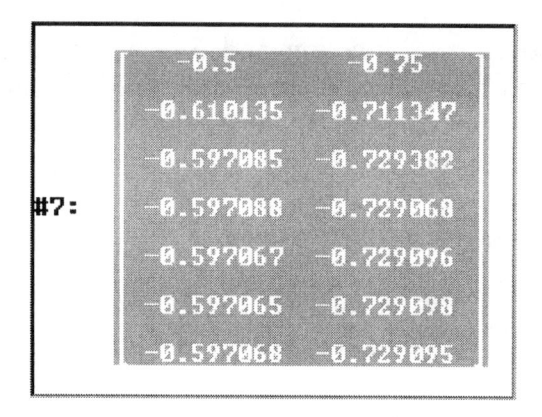

The last two rows indicate that $(-.597068, -.729095)$ is a critical point. If we use the **NEWTONS** command with the other initial guesses, then we find that the other critical points occur at $(.023479, 1.07379)$, $(.527469, -.188337)$ and $(-.59617, .606149)$. Finally, we can classify these critical points by following the method employed in the example above. We start by using the **DIF** command to form the function

$$TEST(x,y) = \left[\frac{\partial^2 f(x,y)}{\partial x^2}, \frac{\partial^2 f(x,y)}{\partial x^2} \frac{\partial^2 f(x,y)}{\partial y^2} - \left(\frac{\partial^2 f(x,y)}{\partial x \partial y} \right)^2 \right].$$

Then we evaluate this function at each of the critical points. The results are shown below.

#18: TEST(-0.597068, -0.729095)
#19: [-2.61744, 1.95076]
#20: TEST(0.023479, 1.07379)
#21: [-0.993126, 3.01639]
#22: TEST(0.527469, -0.188337)
#23: [1.43000, 2.79534]
#24: TEST(-0.59617, 0.606149)
#25: [-1.32222, -2.73947]

From these calculations, we can conclude that $(-.597068, -.729095)$ is a local maximum, $(.023479, 1.07379)$ is a local maximum, $(.527469, -.188337)$ is a local minimum, and $(-.59617, .606149)$ is a saddle point.

Example 38 *As a final example, let's consider the problem of maximizing and minimizing*

$$f(x,y) = x^4 - 3xy + 2y^2$$

on the region

$$S = \{(x,y) \mid x^2 + y^2 \leq 1\}.$$

Since f is a continuous function and S is a closed and bounded set, we know that f has both a maximum value and a minimum value on S. Furthermore, we can find the points at which the maximum and minimum values occur by locating all critical points in the interior of S; i.e. in the set

$$\{(x,y) \mid x^2 + y^2 < 1\}.$$

Then we compare the function values at these points with the maximum and minimum values of f on the boundary of S; i.e. on

$$\{(x,y) \mid x^2 + y^2 = 1\}.$$

From Example 7, we know that the critical points are given by $(3/4, 9/16)$, $(0,0)$ and $(-3/4, -9/16)$. The corresponding function values are shown below.

```
#1:  F(x, y) := x^4 - 3·x·y + 2·y^2

#2:  F(0, 0) = 0

#3:  F(3/4, 9/16) = -81/256

#4:  F(-3/4, -9/16) = -81/256
```

Now we need to maximize and minimize f on the boundary of S. This corresponds to maximizing and minimizing f subject to the constraint $x^2 + y^2 = 1$. We can do this in a couple of ways. One method involves Lagrange multipliers. For this method, we define $g(x,y) = x^2 + y^2$ and then we search for solutions to the system

$$\begin{aligned}\nabla f(x,y) &= s\nabla g(x,y) \\ g(x,y) &= 1\end{aligned}.$$

That is, solutions to the system

$$4x^3 - 3y = 2sx, \quad 4y - 3x = 2sy, \quad x^2 + y^2 = 1.$$

We form this system below.

5.5. OPTIMIZATION AND LAGRANGE MULTIPLIERS

#5: $\left[\dfrac{d}{dx} F(x, y) = 2 \cdot s \cdot x, \; \dfrac{d}{dy} F(x, y) = 2 \cdot s \cdot y, \; x^2 + y^2 = 1 \right]$

#6: $\left[4 \cdot x^3 - 3 \cdot y = 2 \cdot s \cdot x, \; 4 \cdot y - 3 \cdot x = 2 \cdot s \cdot y, \; x^2 + y^2 = 1 \right]$

Then we solve the first equation for s so that we can substitute into the other two equations.

#10: $\mathrm{SOLVE}(4 \cdot x^3 - 3 \cdot y = 2 \cdot s \cdot x, \; s)$

#11: $\left[s = \dfrac{4 \cdot x^3 - 3 \cdot y}{2 \cdot x} \right]$

These equations are rewritten below as expressions which must be set equal to zero.

#12: $s := \dfrac{4 \cdot x^3 - 3 \cdot y}{2 \cdot x}$

#15: $4 \cdot y - 3 \cdot x - \dfrac{y \cdot (4 \cdot x^3 - 3 \cdot y)}{x}, \; x^2 + y^2 - 1$

The implicit plot of the corresponding two equations is shown below.

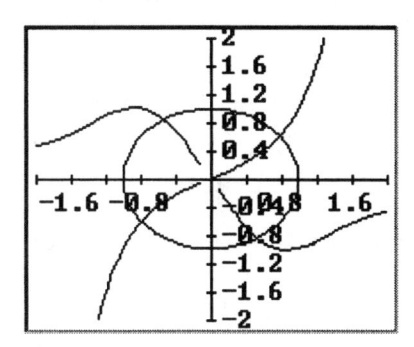

As a result, there must be four solutions. We find the first of these by **Loading** the **Solve.mth** utility file and using the **NEWTONS** command in the form

$$\text{NEWTONS}(\#15, [x, y], [.5, .5], 6)$$

followed by clicking on the [≈] icon.

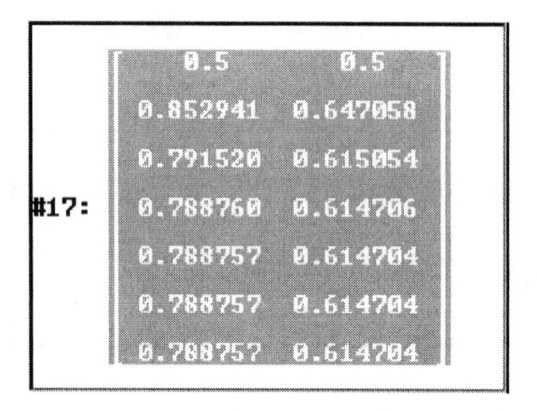

#17:
$$\begin{bmatrix} 0.5 & 0.5 \\ 0.852941 & 0.647058 \\ 0.791520 & 0.615054 \\ 0.788760 & 0.614706 \\ 0.788757 & 0.614704 \\ 0.788757 & 0.614704 \\ 0.788757 & 0.614704 \end{bmatrix}$$

From the values in the last two rows, we can say that one of the solutions is given by $(.788757, .614704)$. If we repeat this process with other initial guesses (taken from the graph above), then we find the other solutions given by $(.533678, -.845688)$, $(-.533678, .845688)$ and $(-.788757, -.614705)$. The corresponding function values are shown below.

#18: F(0.0788757, 0.614704)
#19: 0.610305
#20: F(0.533578, -0.845688)
#21: 2.86515

#22: F(-0.533578, 0.845688)
#23: 2.86515
#24: F(-0.78757, -0.614705)
#25: -0.311914

Comparing these values with the function values at the critical points above, we see that the maximum value of f on S is 2.86547, and it occurs at $(.533678, -.845688)$ and $(-.533678, .845688)$. The minimum value of f on S is $-81/256 \approx -.31641$, and it occurs at both $(3/4, 9/16)$ and $(-3/4, -9/16)$.

Remark: We mentioned above that there were a couple of ways to maximize and minimize

$$f(x, y) = x^4 - 3xy + 2y^2$$

5.5. OPTIMIZATION AND LAGRANGE MULTIPLIERS

subject to the constraint $x^2 + y^2 = 1$. One other way to do this is to parameterize the circle and then evaluate f along the circle. A parameterization for the circle is given by $(\cos(t), \sin(t))$ for $0 \leq t \leq 2\pi$. Consequently, we can maximize and minimize f along the circle by maximizing and minimizing the function

$$g(t) = f(\cos(t), \sin(t)) = \cos^4(t) - 3\cos(t)\sin(t) + 2\sin^2(t)$$

for $0 \leq t \leq 2\pi$.

Exercises:

1. Let $f(x, y) = x^4 - 8x^3 + 25x^2 - 32x - 6xy + 9y^2$.

   (a) Find all critical points of f.

   (b) Produce a graph of $z = f(x, y)$ showing whether the function attains a local maximum, a local minimum or a saddle at each critical point.

   (c) Use the second derivative test to confirm what is indicated by the graph in part b.

   (d) Create a contour plot about each critical point (or all critical points) that substantiates the conclusions of parts b and c.

2. Repeat exercise 1 for the function $f(x, y) = x^4 - 3x^2 - 6xy + 9y^2$.

3. Repeat exercise 1 for the function $f(x, y) = x - x^3 - 3xy^2 + y$.

4. Use a contour plot to find approximate locations for the critical points of

   $$f(x, y) = \left(x^2 - xy + y^2 - y\right) e^{-\left(x^2 + y^2\right)}$$

   for $-2 \leq x \leq 2$ and $-2 \leq y \leq 2$. Then use the **NEWTONS** command to find each of the critical points. Finally, use the second derivative test to determine whether the function attains a local maximum, a local minimum or a saddle at each critical point.

5. Repeat exercise 4 for the function

   $$f(x, y) = \left(x^2 + xy + y^2 - x\right) \cos\left(\pi \sqrt{x^2 + y^2 + 1}\right).$$

6. Use the Lagrange Multipliers method to find the maximum and minimum values of each of the functions in exercises 1-5 on the circle $x^2 + y^2 = 1$.

7. Use the Lagrange Multipliers method to find the maximum and minimum values of each of the functions in exercises 1-5 on the ellipse $2x^2 + 3y^2 = 1$.

8. Use Lagrange multipliers to find the minimum value of

$$f(x, y, z, w) = 4x^2 + 3y^2 + 2z^2 + w^2$$

subject to the constraint

$$x + 2y + 3z + 4w = 100.$$

9. Estimate the minimum value of $f(x, y) = x^2 + y^2$ on the curve

$$x^3 - 2x - 2y^3 + 3y = 0$$

by creating an implicit plot of $x^3 - 2x - 2y^3 + 3y = 0$ and making a geometric observation.

10. Find the minimum value of $f(x, y) = x^2 - x - y + y^2$ on the curve

$$x^3 - 2x - 2y^3 + 3y = 0.$$

11. Find the maximum value of $f(x, y) = x^2 - x - y + y^2$ on the curve

$$x^3 - 2x - 2y^3 + 3y = 0.$$

12. Find the maximum value of $f(x, y) = x^2 - x - y + y^2 - 3x + 7y$ in the region

$$-1 \leq x \leq 1, \ -1 \leq y \leq 1.$$

13. Find all of the points (x, y) at which $f(x, y) = x^2 + y^2$ is minimized, subject to the constraint that there exits a point $(u, v) \neq (x, y)$ so that the normal line to the graph of $z = f(x, y)$ at the point $(u, v, f(u, v))$ passes through $(x, y, f(x, y))$.

Chapter 6

Multiple Integrals

DERIVE has a very impressive integration package which is capable of computing integrals both symbolically and numerically. In this chapter, we will use **DERIVE** to explore double integrals of functions of two variables and triple integrals of functions of three variables.

- Multiple integrals and their applications are the subject of Chapter 12 of Stewart's **Multivariable Calculus: Concepts and Contexts**.

6.1 Double Integrals

- A formal discussion of double integrals can be found in Sections 12.1-12.3 of Stewart's **Multivariable Calculus: Concepts and Contexts**.

Approximating Integrals. Let's suppose we want to approximate the integral of $f(x,y)$ over the rectangle given by $a \leq x \leq b$ and $c \leq y \leq d$ by using a Riemann sum with m equally spaced divisions in the x direction, n equally space directions in the y direction, and function values taken at the midpoint of each rectangular sector. The approximation is given by the formula

$$\sum_{i=1}^{m}\sum_{j=1}^{n} f\left(a + \frac{\Delta x}{2}(2i-1), c + \frac{\Delta y}{2}(2j-1)\right) \Delta x \Delta y$$

where $\Delta x = (b-a)/m$ and $\Delta y = (d-c)/n$. In general, this approximation is referred to as the **midpoint approximation** of the integral.

Example 39 *Let's give the midpoint approximation in the case when the function $f(x,y)$ given by*

$$f(x,y) = \frac{1}{x^2 + 2y^2 + 1},$$

the rectangle is given by $0 \leq x \leq 2$ and $1 \leq y \leq 2$, and $m = n = 10$.

Note that these values imply $\Delta x = 1/5$ and $\Delta y = 1/10$. First, we define the function $f(x,y)$.

#1: $F(x, y) := \dfrac{1}{x^2 + 2 \cdot y^2 + 1}$

Then we author the expression below and click the ≈ icon.

$$\text{SUM}(\text{SUM}(f((2i-1)/10, 1+(2j-1)/10)/50, i, 1, 10), j, 1, 10)$$

#2: $\displaystyle\sum_{j=1}^{10} \sum_{i=1}^{10} \dfrac{F\left(\dfrac{2 \cdot i - 1}{10},\ 1 + \dfrac{2 \cdot j - 1}{20}\right)}{50}$

#3: 0.315608

This computation is repeated below for $m = n = 15$.

#4: $\displaystyle\sum_{j=1}^{15} \sum_{i=1}^{15} \dfrac{F\left(\dfrac{2 \cdot i - 1}{15},\ 1 + \dfrac{2 \cdot j - 1}{30}\right) \cdot 2}{225}$

#5: 0.315649

These calculations indicated that the value of

$$\int_0^2 \int_1^2 \frac{1}{x^2 + 2y^2 + 1} \, dy \, dx$$

6.1. DOUBLE INTEGRALS

should be approximately 0.316. The actual integral can be computed in **DERIVE** by authoring the expression

$$\text{INT}(\text{INT}(f(x,y),x,0,2),y,1,2)$$

and clicking on the [≈] icon.

```
#6:   ∫₁² ∫₀² F(x, y) dx dy

#7:   0.315682
```

Apparently the approximations above are very good.

Symbolic Integration. **DERIVE** has the ability to perform most of the integrations which appear in a standard calculus text.

Example 40 *Let's consider the integral of the function*

$$f(x,y) = x^3 y^2 + xy$$

over the rectangle given by $0 \leq x \leq 1$ and $1 \leq y \leq 2$.

This integral can be computed by evaluating either of the following iterated integrals:

$$\int_0^1 \int_1^2 f(x,y) dy dx \quad \text{or} \quad \int_1^2 \int_0^1 f(x,y) dx dy.$$

We illustrate this point below. First we define the function f.

```
#1:   F(x, y) := x³ y² + x·y
```

Then we author the expression

$$\text{INT}(\text{INT}(f(x,y),x,0,1),y,1,2)$$

and press the ■ icon, and repeat the process with the **Expression**

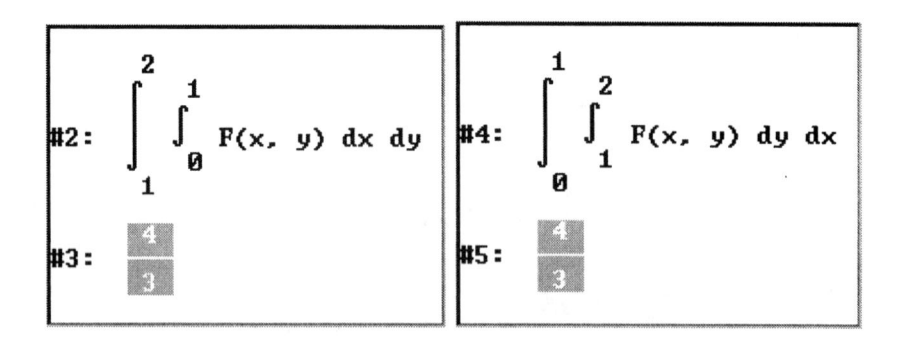

$$INT(INT(f(x,y),y,1,2),x,0,1).$$

#2: $\int_1^2 \int_0^1 F(x, y) \, dx \, dy$ #4: $\int_0^1 \int_1^2 F(x, y) \, dy \, dx$

#3: $\dfrac{4}{3}$ #5: $\dfrac{4}{3}$

In this case, the value 4/3 represents a volume since the function $f(x,y)$ is nonnegative over the rectangle. More specifically, it is the volume of the solid bounded between $z = 0$ and $z = f(x, y)$ for $0 \le x \le 1$ and $1 \le y \le 2$.

Example 41 *Now let's find the exact volume of the solid which lies under the surface $z = xy^2$ over the triangular region described by $0 \le x \le 1$ and $x \le y \le 2x$.*

The triangular region between the graphs of $y = x$ and $y = 2x$ is shown below by authoring the expression

$$\mathbf{VECTOR}([n*x], n, 1, 2, 1/10)$$

and clicking on the ■ icon.

#1: $VECTOR\left([n \cdot x], n, 1, 2, \dfrac{1}{10}\right)$

Then the matrix which is created is graphed in a **2-D Plot** window for $0 \le x \le 1$.

6.1. DOUBLE INTEGRALS

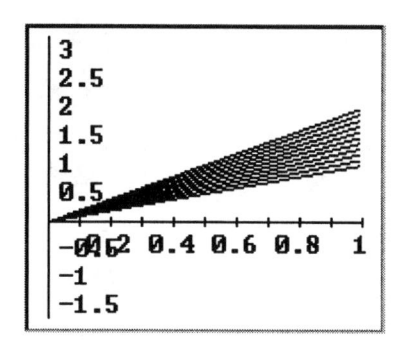

The volume can be computed by evaluating the iterated integral

$$\int_0^1 \int_x^{2x} f(x,y) \, dy \, dx.$$

We simply define the function f, author the expression

$$\text{INT}(\text{INT}(f(x,y), y, x, 2*x), x, 0, 1)$$

and click on the ▣ icon.

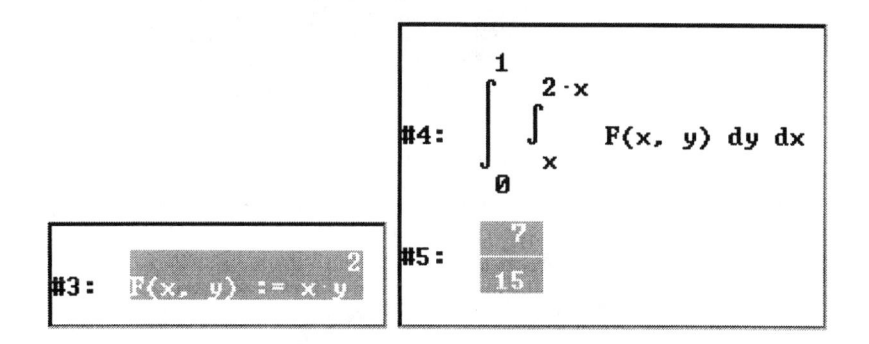

This volume can also be computed by reversing the order of the iterated integral, but the process is more complicated. In this case we obtain the volume as the sum of two integrals given by

$$\int_0^1 \int_{y/2}^y f(x,y) \, dx \, dy + \int_1^2 \int_{y/2}^1 f(x,y) \, dx \, dy.$$

As we can see below, the value is the same.

#6: $\displaystyle\int_0^1 \int_{y/2}^{y} F(x, y)\, dx\, dy + \int_1^2 \int_{y/2}^{1} F(x, y)\, dx\, dy$

#7: $\dfrac{7}{15}$

Example 42 *Finally, let's find the exact volume of the solid that lies under the surface $z = x^2 y^2$ and over the region described by $0 \leq x \leq \sin(y)$ and $0 \leq y \leq \pi$.*

This region is graphed and shaded below in **DERIVE** by authoring the expression

$$\text{VECTOR}([n * \sin(t), t], n, 0, 1, 1/10)$$

and clicking on the ▨ icon to create a matrix of parameterized curves.

#1: $\text{VECTOR}\left([n \cdot \text{SIN}(t), t], n, 0, 1, \dfrac{1}{20}\right)$

The result is plotted in a **2-D Plot** window.

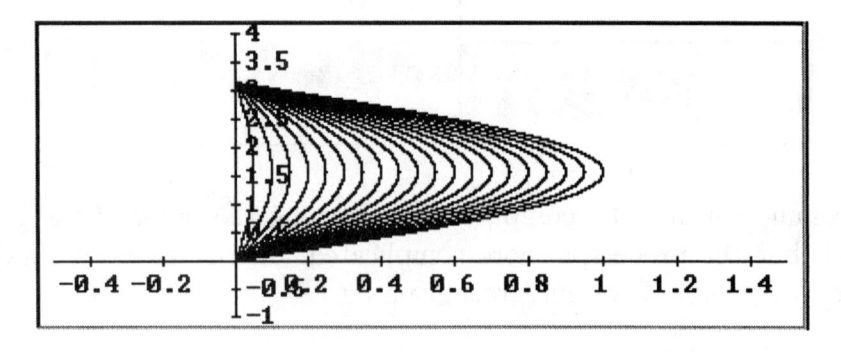

The volume can be computed by evaluating the iterated integral

$$\int_0^\pi \int_0^{\sin(y)} x^2 y^2 \, dx\, dy.$$

6.2. POLAR COORDINATES

It can also be computed by evaluating the iterated integral

$$\int_0^1 \int_{\sin^{-1}(x)}^{\pi - \sin^{-1}(x)} x^2 y^2 \, dy \, dx.$$

Both integrations are given below along with the numerical approximation of the resulting value. Note that we have defined (but not shown) $\mathbf{f(x,y)} := \mathbf{x^2 y^2}$.

```
#2:    ∫₀^π ∫₀^SIN(y) F(x, y) dx dy

#3:    1.20559
```

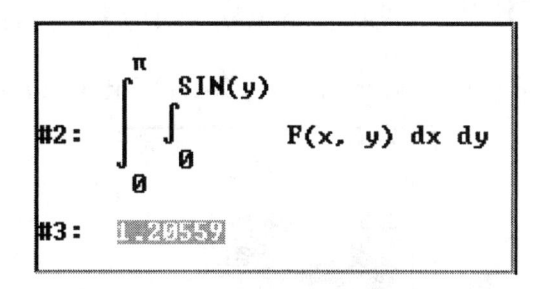

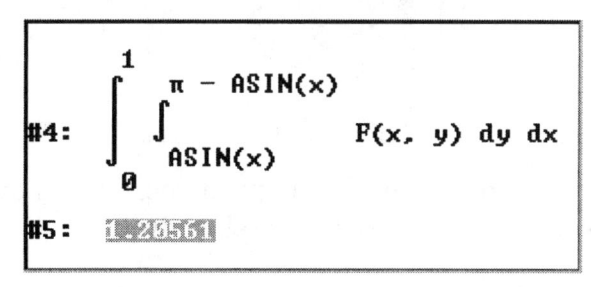

```
#4:    ∫₀^1 ∫_{ASIN(x)}^{π - ASIN(x)} F(x, y) dy dx

#5:    1.20561
```

Consequently, the volume of the solid is roughly 1.2056 cubic units.

6.2 Polar Coordinates

- Double integrals of functions expressed in polar coordinates are discussed in Section 12.4 of Stewart's **Multivariable Calculus: Concepts and Contexts**. Stewart shows that if $f(x,y)$ is continuous over a region D in the x,y-plane described in polar coordinates by

$$g(\theta) \leq r \leq h(\theta) \text{ and } a \leq \theta \leq b$$

then the integral of $f(x,y)$ over D is given by

$$\int_a^b \int_{g(\theta)}^{h(\theta)} f(r\cos(\theta), r\sin(\theta)) \, r \, dr \, d\theta.$$

Example 43 *Let's use polar coordinates to find the integral of*

$$f(x,y) = \frac{1}{2}(x-y)^2$$

over the sector of the unit disk in which $-\pi/4 \le \theta \le \pi/4$.

The sector shown below is created by authoring the expression

VECTOR([t * cos(s), t * sin(s)], s, ipi/4, pi/4, pi/20),

clicking on the ≈ icon, and creating a **2-D Plot** window with $0 \le t \le 1$.

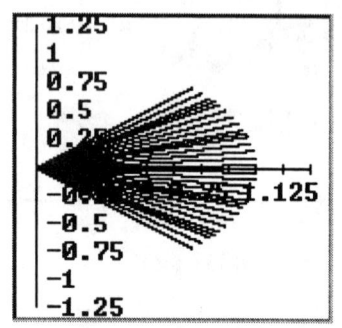

Note that this sector can be described in polar coordinates by $0 \le r \le 1$ and $-\pi/4 \le \theta \le \pi/4$. Consequently, the integral is given by

$$\int_{-\pi/4}^{\pi/4} \int_0^1 f(r\cos(\theta), r\sin(\theta))\, r\, dr\, d\theta.$$

The computation is performed by defining the function f, authoring the expression

INT(INT(f(r * cos(s), r * sin(s)) * r, r, 0, 1), s, -pi/4, pi/4)

and clicking on the = icon.

#6: $\displaystyle\int_0^1 \int_{-\pi/4}^{\pi/4} F(r \cdot COS(s), r \cdot SIN(s)) \cdot r\ ds\ dr$

#7: $\dfrac{\pi}{16}$

6.2. POLAR COORDINATES

Example 44 *Finally, let's use polar coordinates to compute the volume of the solid that lies under the surface*

$$z = x + y^2$$

and above the disk bounded by the circle

$$x^2 + (y-1)^2 = 1.$$

To start, we need a polar description of the disk. The necessary calculations are shown below. First we define a function which helps us define the disk and write the equation above in polar coordinates.

```
#1:  G(x, y) := x^2 + (y - 1)^2
#2:  G(r·COS(θ), r·SIN(θ)) = 1
#3:  - 2·r·SIN(θ) + r^2 + 1 = 1
```

Then we use the **SOLVE** command to determine r.

```
#4:  SOLVE(- 2·r·SIN(θ) + r^2 + 1 = 1, r)
#5:  [r = 0, r = 2·SIN(θ)]
```

Consequently, the disk is described by $0 \leq r \leq 2\sin(\theta)$ and $0 \leq \theta \leq \pi$. As a result, the volume is given by the double integral

$$\int_0^\pi \int_0^{2\sin(\theta)} f(r\cos(\theta), r\sin(\theta))\, r\, dr\, d\theta$$

where $f(x, y) = x + y^2$.

```
#6:  F(x, y) := x + y^2
```

The double integral is computed below.

#7: $\displaystyle\int_0^{\pi}\int_0^{2\cdot\text{SIN}(\theta)} F(r\cdot\text{COS}(\theta), r\cdot\text{SIN}(\theta))\cdot r\ dr\ d\theta$

#8: $\dfrac{5\cdot\pi}{4}$

6.3 Applications

- Sections 12.5 and 12.6 of Stewart's **Multivariable Calculus: Concepts and Contexts** deal with applications of double integrals. The important concepts from Stewart's sections are mass, moments, centers of mass and surface area.

Example 45 *Let's find the center of mass of a lamina with mass density $\rho(x,y) = 3x + y^2$ which occupies the region in the first quadrant bounded by the graphs of $y = x^2$ and $x = 1$.*

The region shown below is obtained by authoring the expression

$$\text{VECTOR}([n*x\hat{\ }2], n, 0, 1, 1/20),$$

clicking on the ▇ icon, and graphing the resulting matrix in a **2-D Plot** window.

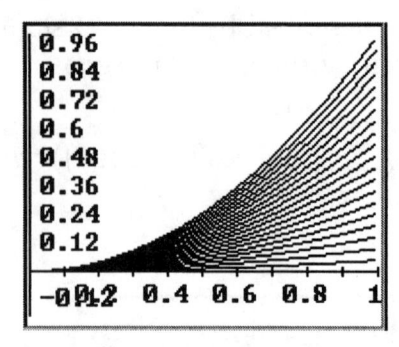

The mass is given by

$$m = \int_0^1 \int_0^{x^2} \rho(x,y)\,dy\,dx$$

6.3. APPLICATIONS

and the moments about the x and y axes are given respectively by

$$M_x = \int_0^1 \int_0^{x^2} y\rho(x,y)dydx$$

and

$$M_y = \int_0^1 \int_0^{x^2} x\rho(x,y)dydx.$$

These values are computed below using the definition

$$\mathbf{p(x,y) := 3x + y^2}.$$

#1: $P(x, y) := 3 \cdot x + y^2$

#2: $m := \int_0^1 \int_0^{x^2} P(x, y) \, dy \, dx = \dfrac{67}{84}$

#3: $mx := \int_0^1 \int_0^{x^2} y \cdot P(x, y) \, dy \, dx = \dfrac{5}{18}$

#4: $my := \int_0^1 \int_0^{x^2} x \cdot P(x, y) \, dy \, dx = \dfrac{77}{120}$

The values above can be used to compute the center of mass $(\bar{x}, \bar{y})$ by using the formula

$$(\bar{x}, \bar{y}) = \left(\frac{M_y}{m}, \frac{M_x}{m}\right).$$

The center of mass is given below.

```
#5:  my/m   mx/m  =  539/670   70/201
```

Example 46 *Let's find the surface area of the portion of the graph of*

$$z = 1 - 4x^2y^2$$

associated with values (x, y) taken from the unit disk.

Recall that the surface area of the graph of $z = f(x, y)$ over a region D in the xy-plane is given by

$$\iint_D \sqrt{f_x(x,y)^2 + f_y(x,y)^2 + 1}\, dA.$$

In our setting, it will be easier to work with polar coordinates. We have

$$f(x, y) = 1 - 4x^2y^2$$

and hence

$$f_x(x, y) = -8xy^2 \text{ and } f_y(x, y) = -8x^2y.$$

Also, the unit disk is given in polar coordinates by $0 \leq r \leq 1$ and $0 \leq \theta \leq 2\pi$. The necessary calculations can be performed by first defining f and computing its first order partials.

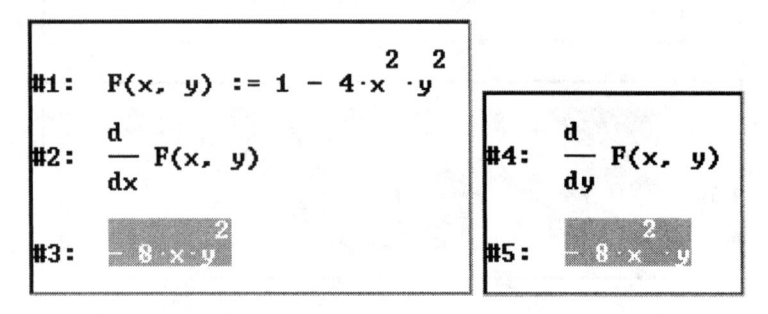

```
#1:  F(x, y) := 1 - 4·x²·y²

#2:  d/dx F(x, y)              #4:  d/dy F(x, y)

#3:  -8·x·y²                   #5:  -8·x²·y
```

Then we define the integrand for the surface integral.

```
#6:  G(x, y) := √((-8·x·y²)² + (-8·x²·y)² + 1)
```

6.3. APPLICATIONS

Finally, we compute the integral by using the ≈ icon. **Note:** This calculation takes several minutes.

```
#7:    ∫₀^(2·π) ∫₀^1 G(r·COS(θ), r·SIN(θ))·r dr dθ

#8:    [result]
```

Exercises:

1. Use successively refined midpoint approximations to compute each of the following double integrals to two significant digits. Compare your results with the numerical approximation given by **DERIVE** for each of the double integrals.

    (a) $\int_0^{\pi/2} \int_0^\pi \sqrt{\cos(x) + \sin(y)}\, dx\, dy$

    (b) $\int_0^2 \int_0^3 \frac{1}{1+x^2+y^2}\, dx\, dy$

    (c) $\int_0^1 \int_0^1 \exp\left(-\left(x^2 + y^2\right)\right) dx\, dy$

2. Compute the integral of $f(x,y) = x^2 + y^2$ over each of the following regions.

    (a) the triangle with vertices $(0,0)$, $(1,0)$, and $(1,1)$.
    (b) the triangle with vertices $(0,0)$, $(0,1)$, and $(1,0)$.
    (c) the region between $y = 0$ and $y = \sin(x)$ for $0 \le x \le \pi$.
    (d) the unit disk.

3. Compute the volume of the solid that lies under the surface
$$z = \cos\left(\frac{\pi}{2}\left(x^2 + y^2\right)\right)$$
and above the unit disk in the xy-plane. Also, graph $z = \cos\left(\frac{\pi}{2}\left(x^2 + y^2\right)\right)$ and use the graph to make a sketch of the solid.

4. Compute the volume of the solid that lies under the surface
$$z = \sin(x)\sin(y)$$
and above the square described by $0 \leq x \leq \pi$ and $0 \leq y \leq \pi$. Also, give a sketch of the solid.

5. Find the center of mass of the lamina that lies within the unit circle and below the line $x + y = 1$, assuming that the mass density is given by $\rho(x, y) = 2$.

6. Find the center of mass of the lamina that occupies the region inside the circle $x^2 + (y - 1)^2 = 1$ and outside the circle $x^2 + y^2 = 1$, given that the mass density is given by $\rho(x, y) = x + y + 1$.

7. Approximate the area of the portion of the surface $z = x + y^2$ that lies above the unit disk.

8. Approximate the area of the surface $z = x^2 y^2$ that lies above the unit disk.

9. Approximate the area of the portion of the surface $z = x^2 y^2$ that lies above the square described by $-1 \leq x \leq 1$ and $-1 \leq y \leq 1$.

6.4 Triple Integrals

- Triple integrals are defined and developed in Section 12.7 of Stewart's **Multivariable Calculus: Concepts and Contexts**.

We demonstrate the use of **DERIVE** in this section to compute triple integrals. Our experience from the last example from the previous section should tell us that numerical approximation of triple integrals will be difficult. Luckily, **DERIVE** can compute many triple integrals exactly.

Example 47 *Let's begin by computing the integral of*
$$f(x, y, z) = z\sqrt{xy}$$
over the region that lies under the paraboloid $z = x^2 + y^2$ and above the triangle in the xy-plane bounded by the coordinate axes and the line $x + y = 1$.

6.4. TRIPLE INTEGRALS

The triangle in the xy-plane is obtained by authoring the expression

$$\mathbf{VECTOR}([s, t*(1-s)], s, 0, 1, 1/40),$$

clicking on the [=] icon, and graphing the result in a **2-D Plot** window.

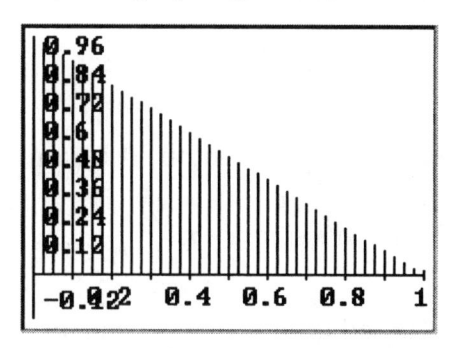

Consequently, we can perform the triple integral over the region by noting that it can be described by

$$0 \leq z \leq x^2 + y^2, \ 0 \leq y \leq 1-x, \ 0 \leq x \leq 1.$$

This leads to the triple integral

$$\int_0^1 \int_0^{1-x} \int_0^{x^2+y^2} z\sqrt{xy}\, dz\, dy\, dx.$$

We compute this integral below.

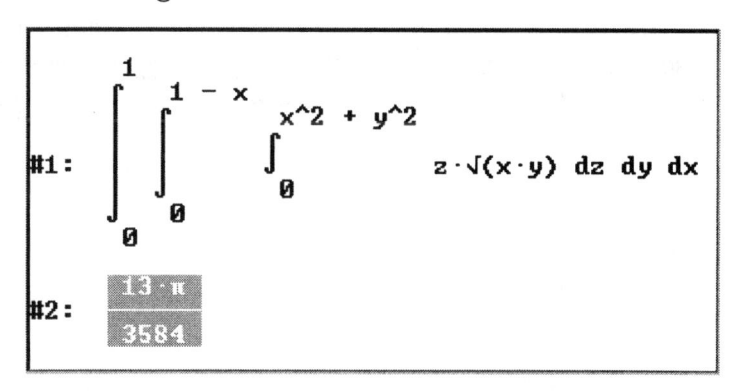

#1: $\int_0^1 \int_0^{1-x} \int_0^{x^2+y^2} z \cdot \sqrt{(x \cdot y)}\, dz\, dy\, dx$

#2: $\dfrac{13 \cdot \pi}{3584}$

Example 48 *Find the volume and mass of a solid occupies the region that lies under the plane*

$$z = x - y/4$$

and above the planar region between the graphs of

$$y = \pm \sin(\pi x) \text{ for } 0 \le x \le 1,$$

assuming that the mass density is given by

$$\rho(x, y, z) = 1 + y^2 + z^2.$$

The planar region shown below is obtained by authoring the expression

VECTOR$([n * \sin(pi * x)], n, -1, 1, 1/20),$

clicking on the ■ icon, and graphing the result in a **2-D Plot** window.

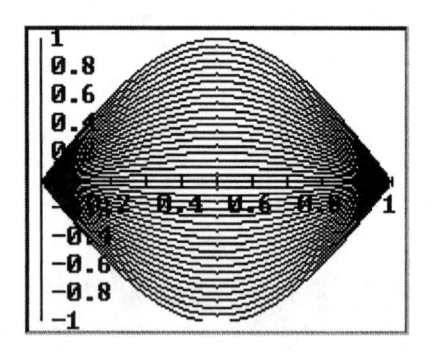

Now, it is not hard to see that $x - y/4 \ge 0$ over this planar region. Consequently, the solid can be described by $0 \le z \le x - y/4$, $-\sin(\pi x) \le y \le \sin(\pi x)$ and $0 \le x \le 1$. Consequently, the volume of the solid is given by

$$\int_0^1 \int_{-\sin(\pi x)}^{\sin(\pi x)} \int_0^{x-y/4} dz\, dy\, dx$$

and the mass of the solid is given by

$$\int_0^1 \int_{-\sin(\pi x)}^{\sin(\pi x)} \int_0^{x-y/4} \rho(x, y, z)\, dz\, dy\, dx.$$

Consequently, the volume is given by

6.4. TRIPLE INTEGRALS

#2: $\displaystyle\int_{0}^{1}\int_{-\sin(\pi\cdot x)}^{\sin(\pi\cdot x)}\int_{0}^{x-y/4} 1 \, dz \, dy \, dx$

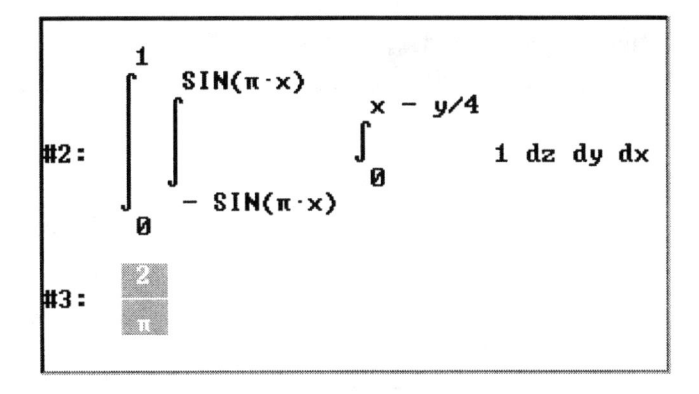

#3: $\dfrac{2}{\pi}$

and the mass is given by first defining the density

#1: $P(x, y, z) := 1 + y^2 + z^2$

and then computing the integral below.

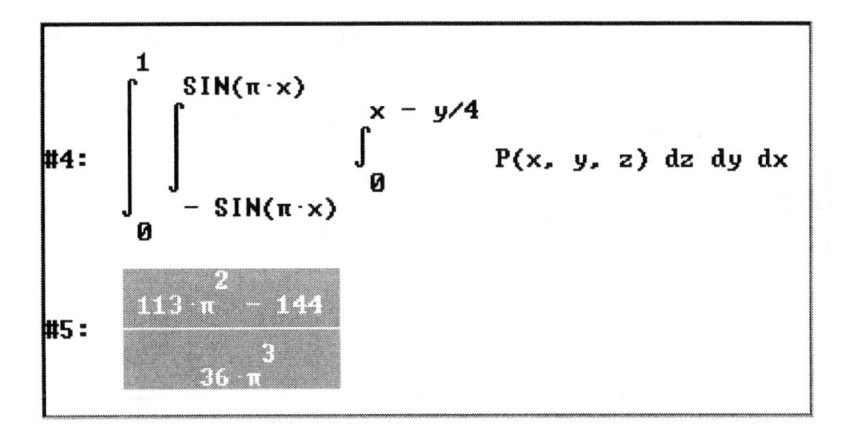

#4: $\displaystyle\int_{0}^{1}\int_{-\sin(\pi\cdot x)}^{\sin(\pi\cdot x)}\int_{0}^{x-y/4} P(x, y, z) \, dz \, dy \, dx$

#5: $\dfrac{113\pi^2 - 144}{36\pi^3}$

Example 49 *Let's find the center of mass of the solid in the example above.*

We start by storing the mass from the calculation above as m

#6: $m := \dfrac{113\pi^2 - 144}{36\pi^3}$

and computing the moment M_{yz}.

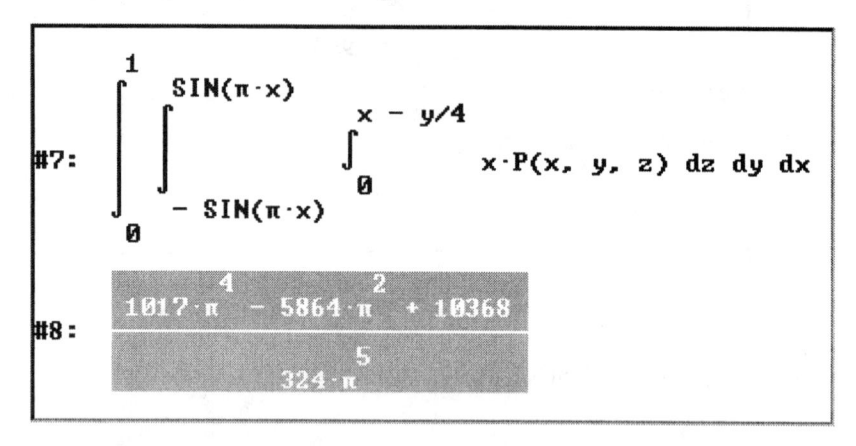

#7: $\displaystyle\int_{0}^{1}\int_{-\sin(\pi\cdot x)}^{\sin(\pi\cdot x)}\int_{0}^{x-y/4} x\cdot P(x,y,z)\,dz\,dy\,dx$

#8: $\dfrac{1017\cdot\pi^{4}-5864\cdot\pi^{2}+10368}{324\cdot\pi^{5}}$

In a similar manner, we can compute M_{xz}

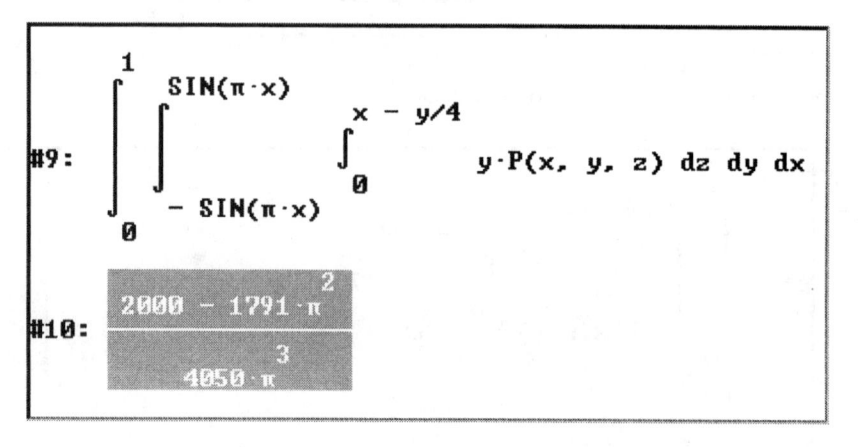

#9: $\displaystyle\int_{0}^{1}\int_{-\sin(\pi\cdot x)}^{\sin(\pi\cdot x)}\int_{0}^{x-y/4} y\cdot P(x,y,z)\,dz\,dy\,dx$

#10: $\dfrac{2000-1791\cdot\pi^{2}}{4050\cdot\pi^{3}}$

and finally M_{xy}

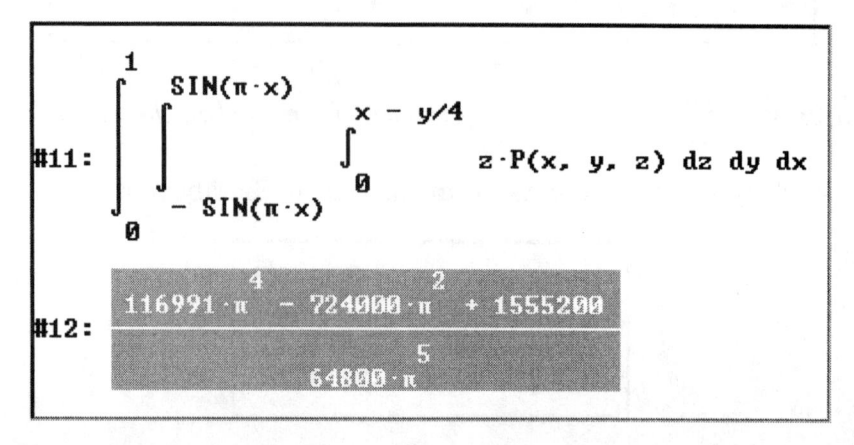

#11: $\displaystyle\int_{0}^{1}\int_{-\sin(\pi\cdot x)}^{\sin(\pi\cdot x)}\int_{0}^{x-y/4} z\cdot P(x,y,z)\,dz\,dy\,dx$

#12: $\dfrac{116991\cdot\pi^{4}-724000\cdot\pi^{2}+1555200}{64800\cdot\pi^{5}}$

Consequently, since the center of mass is computed using the formula

$$(\bar{x}, \bar{y}, \bar{z}) = \left(\frac{M_{yz}}{m}, \frac{M_{xz}}{m}, \frac{M_{xy}}{m} \right)$$

the center of mass is approximately $(.597604, -.143469, .336462)$.

6.5 Cylindrical and Spherical Coordinates

- Triple integrals in cylindrical and spherical coordinates are discussed in Section 12.8 of Stewart's **Multivariable Calculus: Concepts and Contexts**. Also, see the discussion in Section 3.3 of this manual.

Knowing how to compute a triple integral with **DERIVE** is sufficient for computing integrals in either cylindrical or spherical coordinates. For this reason, we only show the computations in the first example below.

Example 50 *(cylindrical coordinates)* Compute the volume and mass of the solid which lies under the surface $z = 1 - 4x^2 y^2$ and over the unit disk in the xy-plane, given that the mass density is given by

$$\rho(x, y, z) = 3 - 2(x^2 + y^2).$$

It is not hard to show that $1 - 4x^2 y^2 \geq 0$ over the unit disk. Consequently, the solid region can be described in cylindrical coordinates via

$$0 \leq \theta \leq 2\pi, \ 0 \leq r \leq 1 \text{ and } 0 \leq z \leq 1 - 4r^4 \cos^2(\theta) \sin^2(\theta).$$

As a result, the volume is given by

$$\int_0^1 \int_0^{2\pi} \int_0^{1-4r^4 \cos^2(\theta) \sin^2(\theta)} r \, dz \, d\theta \, dr.$$

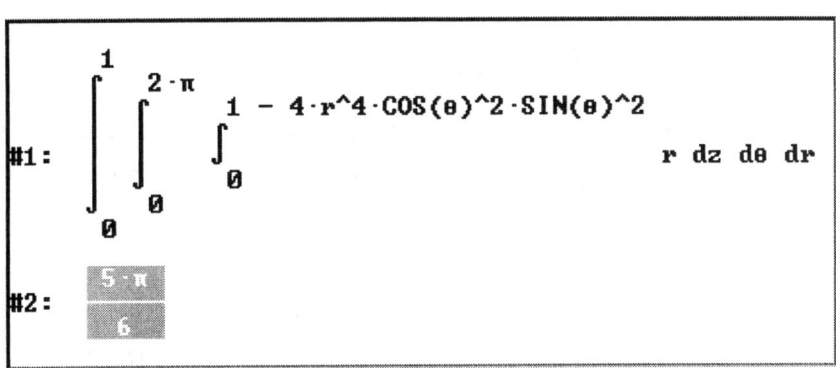

#1: $\displaystyle \int_0^1 \int_0^{2\pi} \int_0^{1 - 4 \cdot r^4 \cdot \text{COS}(\theta)^2 \cdot \text{SIN}(\theta)^2} r \, dz \, d\theta \, dr$

#2: $\displaystyle \frac{5 \cdot \pi}{6}$

Now, recall that the mass density is given by

$$\rho(x, y, z) = 3 - 2\left(x^2 + y^2\right).$$

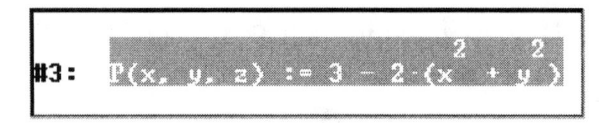

#3: $\rho(x, y, z) := 3 - 2 \cdot (x^2 + y^2)$

So, the mass is given by

$$\int_0^1 \int_0^{2\pi} \int_0^{1-4r^4 \cos^2(\theta) \sin^2(\theta)} \rho(r\cos(\theta), r\sin(\theta), z)\, r\, dz\, d\theta\, dr.$$

This calculation is shown below.

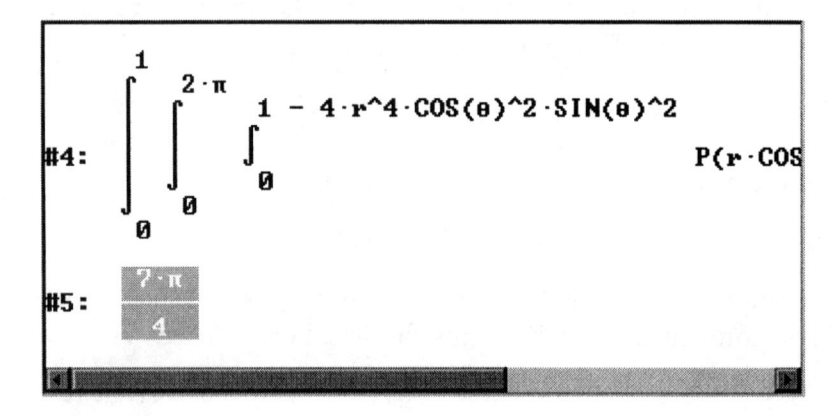

#4: $\int_0^1 \int_0^{2\pi} \int_0^{1 - 4 \cdot r^4 \cdot \cos(\theta)^2 \cdot \sin(\theta)^2} P(r \cdot \cos\ldots$

#5: $\dfrac{7 \cdot \pi}{4}$

Example 51 *(spherical coordinates) The density of a spherical shell occupying the region $2 \le \rho \le 3$ is proportional to the square of the distance from the point $(0, 0, -3)$ and given by*

$$R(x, y, z) = \frac{1}{10}\left(x^2 + y^2 + (z+3)^2\right)$$

(note that we are using R for density since the variable ρ is used in the spherical coordinates we are using to describe our problem). Let's find the mass and center of mass of the solid.

We start by denoting

$$G(\rho, \phi, \theta) = R(\rho\sin(\phi)\cos(\theta), \rho\sin(\phi)\sin(\theta), \rho\cos(\phi))\rho^2 \sin(\phi).$$

6.5. CYLINDRICAL AND SPHERICAL COORDINATES

Then the mass is given by

$$m = \int_0^{2\pi} \int_0^\pi \int_2^3 G(\rho, \phi, \theta) d\rho d\phi d\theta.$$

The value computed in **DERIVE** is $992\pi/25$. Now, because of symmetry, each of the moments M_{yz} and M_{xy} will be zero. Also,

$$M_{xy} = \int_0^{2\pi} \int_0^\pi \int_2^3 \rho \cos(\phi) G(\rho, \phi, \theta) d\rho d\phi d\theta,$$

and **DERIVE** gives this value as $844\pi/25$. Consequently, the center of mass is given by

$$\left(0, 0, \frac{221}{248}\right).$$

Exercises:

1. A solid occupies the region between the paraboloid

$$z = 1 + x^2 + y^2$$

and the plane $x+y+z = 0$ that lies above the square in which $0 \le x \le 1$ and $0 \le y \le 1$. Its mass density is given by

$$\rho(x, y, z) = 2 + z - x^2 - y^2.$$

Find:

(a) the volume of the solid;
(b) the mass of the solid;
(c) the solid's center of mass.

2. A solid occupies the region bounded by the paraboloid

$$z = 1 + x^2 + y^2,$$

the plane $x + y + z = 0$, and the cylinder $x^2 + y^2 = 1$. Its mass density is given by

$$\rho(x, y, z) = 1 + z - x^2 - y^2.$$

Use cylindrical coordinates to find:

(a) the volume of the solid;
(b) the mass of the solid;
(c) the solid's center of mass.

3. A solid occupies the region bounded by the sphere
$$x^2 + y^2 + z^2 = 1$$
and the cone $z = x^2 + y^2$. Its mass density is given by
$$\rho(x, y, z) = 1 + z - x^2 - y^2.$$
Use spherical coordinates to find:

(a) the volume of the solid;
(b) the mass of the solid;
(c) the solid's center of mass.

6.6 Change of Variables

- The change of variables formula for multiple integrals is the subject of Section 12.9 of Stewart's **Multivariable Calculus: Concepts and Contexts**.

Let's start with the two dimensional setting. We know from Stewart's text that if T is a C^1 transformation that maps the region S in the uv-plane onto the region R in the xy-plane then the Jacobian matrix of the transformation T is given by
$$J(u, v) = \begin{pmatrix} g_u & g_v \\ h_u & h_v \end{pmatrix}$$
and the Jacobian determinant of T is given by
$$\frac{\partial(x, y)}{\partial(u, v)} = \det(J(u, v)).$$

DERIVE's built in command **JACOBIAN** automatically computes the Jacobian matrix of T. The **JACOBIAN** command is part of the **Vector.mth** utility file, and consequently this file must be **Loaded** before the command is available.

6.6. CHANGE OF VARIABLES

Example 52 *Compute the Jacobian matrix and determinant of the transformation given by*
$$T(u,v) = \left(u + 3v, u^2 - 2v\right).$$

The computations are shown below. Recall from above that we must **Load** the utility file **Vector.mth** prior to using the **JACOBIAN** command.

#1: $\text{JACOBIAN}(\left[u + 3 \cdot v, u^2 - 2 \cdot v\right], [u, v])$

#2: $\begin{bmatrix} 1 & 3 \\ 2 \cdot u & -2 \end{bmatrix}$

The **DET** command computes the determinant of the Jacobian matrix.

#3: $\text{DET}\begin{bmatrix} 1 & 3 \\ 2 \cdot u & -2 \end{bmatrix} = -6 \cdot u - 2$

In order to put this to use in the setting of integration, we simply have to recall that if T is a one-to-one and onto transformation from the region S in the uv-plane to the region R in the xy-plane and we make the change of variables
$$(x, y) = T(u, v),$$
then
$$\int\int_R f(x,y)\, dx dy = \int\int_S f(T(u,v)) \left|\frac{\partial(x,y)}{\partial(u,v)}\right| du dv.$$

Example 53 *Use a change of variables to compute*
$$\int\int_R \left(x^2 - y\right) dx dy$$
where R is the triangle in the xy-plane with vertices $(1,2)$, $(3,5)$ and $(-1,7)$.

The region R is shown below.

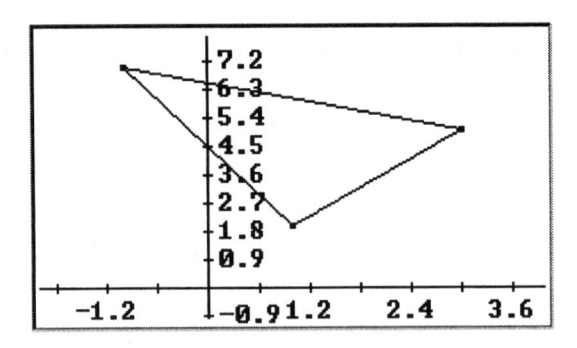

We will see that we can create a simple transformation $T(u,v)$ which maps the triangle with vertices $(0,0)$, $(1,0)$ and $(0,1)$ in the uv-plane to the triangle above. The idea is to look for a map of the form

$$(x,y) = T(u,v) = (au + bv, cu + dv) + (e, f)$$

so that

$$(1,2) = T(0,0), \ (3,5) = T(1,0) \text{ and } (-1,7) = T(0,1).$$

These equations give

$$(1,2) = (e, f), \ (3,5) = (a, c) + (e, f) \text{ and } (-1,7) = (b, d) + (e, f),$$

implying

$$(a, c) = (2, 3) \text{ and } (b, d) = (-2, 5).$$

So,

$$T(u,v) = (2u - 2v + 1, 3u + 5v + 2)$$

and

$$\frac{\partial(x,y)}{\partial(u,v)} = 16.$$

Consequently,

$$\int\int_R (x^2 - y)\, dx dy = \int_0^1 \int_0^{1-u} [(2u - 2v + 1)^2 - (3u + 5v + 2)]\, 16\, dv du$$
$$= -24.$$

Remark: In general, we can define a transformation T which maps the triangle with vertices $(0,0)$, $(1,0)$ and $(0,1)$ in the uv-plane to the triangle with vertices (a_1, b_1), (a_2, b_2) and (a_3, b_3) in the xy-plane with the transformation

$$T(u,v) = ((a_2 - a_1)u + (a_3 - a_1)v + a_1, (b_2 - b_1)u + (b_3 - b_1)v + b_1).$$

6.6. CHANGE OF VARIABLES

Example 54 *Use a change of variables to compute*

$$\int\int_R xy\, dx dy$$

where R is the region in the xy-plane that is bounded by the lines $y = x$ and $y = x - 1$ and the circles of radius $\sqrt{2}$ centered at the points $(\sqrt{2}, \sqrt{2})$ and $(0, -\sqrt{2})$.

We can sketch this region in **DERIVE** by plotting $y = x$ and $y = x - 1$ along with the curves given parametrically by

$$\left[\sqrt{2} + \sqrt{2}\cos(t), \sqrt{2} + \sqrt{2}\sin(t)\right]$$

and

$$\left[\sqrt{2}\cos(t), -\sqrt{2} + \sqrt{2}\sin(t)\right]$$

for $0 \le t \le 2\pi$.

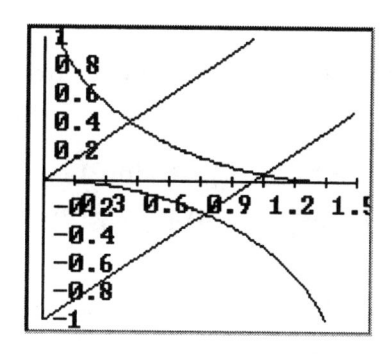

Since the lines $y = x$ and $y = x - 1$ are parallel, we can use a change of variables to compute

$$\int\int_R xy\, dx dy.$$

The idea is to rotate these lines in the xy-plane to the lines $u = 0$ and $u = const$ in the uv-plane. Notice that this represents a counter-clockwise rotation of $\pi/4$ radians. In general, a counter-clockwise rotation of θ radians of a region in the xy-plane to a region in the uv-plane can be achieved via the transformation

$$(x, y) = T(u, v) = (\cos(\theta)u + \sin(\theta)v, -\sin(\theta)u + \cos(\theta)v)$$

Consequently, we need to make the change of variables

$$(x, y) = T(u, v) = \left(\frac{\sqrt{2}}{2}u + \frac{\sqrt{2}}{2}v, -\frac{\sqrt{2}}{2}u + \frac{\sqrt{2}}{2}v\right).$$

Then we can check that $y = x$ corresponds to

$$-\frac{\sqrt{2}}{2}u + \frac{\sqrt{2}}{2}v = \frac{\sqrt{2}}{2}u + \frac{\sqrt{2}}{2}v,$$

implying

$$u = 0.$$

Also, $y = x - 1$ corresponds to

$$-\frac{\sqrt{2}}{2}u + \frac{\sqrt{2}}{2}v = \frac{\sqrt{2}}{2}u + \frac{\sqrt{2}}{2}v - 1,$$

implying

$$u = \frac{\sqrt{2}}{2}.$$

Now let's see what happens to the circles above. In general, we can solve the equations

$$\begin{aligned} x &= \cos(\theta)u + \sin(\theta)v \\ y &= -\sin(\theta)u + \cos(\theta)v \end{aligned}$$

for u and v. We simply author the expression

SOLVE$([x = \cos(\theta)u + \sin(\theta)v, y = -\sin(\theta)u + \cos(\theta)v], [u, v])$

and click on the ▣ icon. This yields

```
#2:   [ u = x·COS(θ) - y·SIN(θ)    v = y·COS(θ) + x·SIN(θ) ]
```

So, in our case, since $\theta = \pi/4$, we have

$$u = \frac{\sqrt{2}}{2}x - \frac{\sqrt{2}}{2}y$$

$$v = \frac{\sqrt{2}}{2}x + \frac{\sqrt{2}}{2}y.$$

6.6. CHANGE OF VARIABLES

We can use these equations to determine what happens with our circles. Note that the circle of radius $\sqrt{2}$ centered at $(\sqrt{2}, \sqrt{2})$ can be parameterized by

$$(x, y) = \left(\sqrt{2} + \sqrt{2}\cos(t), \sqrt{2} + \sqrt{2}\sin(t)\right) \text{ for } 0 \le t \le 2\pi.$$

Substituting this above gives

$$\begin{aligned} u &= \frac{\sqrt{2}}{2}\left(\sqrt{2} + \sqrt{2}\cos(t)\right) - \frac{\sqrt{2}}{2}\left(\sqrt{2} + \sqrt{2}\sin(t)\right) \\ &= \cos(t) - \sin(t). \end{aligned}$$

Similarly,

$$\begin{aligned} v &= \frac{\sqrt{2}}{2}\left(\sqrt{2} + \sqrt{2}\cos(t)\right) + \frac{\sqrt{2}}{2}\left(\sqrt{2} + \sqrt{2}\sin(t)\right) \\ &= 2 + \cos(t) + \sin(t). \end{aligned}$$

Note that

$$u^2 + (v - 2)^2 = 2.$$

That is, the circle of radius $\sqrt{2}$ centered at the point $(\sqrt{2}, \sqrt{2})$ in the xy-plane corresponds to the circle of radius $\sqrt{2}$ centered at the point $(0, 2)$ in the uv-plane. For the other circle, note that the circle of radius $\sqrt{2}$ centered at $(0, -\sqrt{2})$ can be parameterized by

$$(x, y) = \left(\sqrt{2}\cos(t), -\sqrt{2} + \sqrt{2}\sin(t)\right) \text{ for } 0 \le t \le 2\pi.$$

Substituting this above gives

$$\begin{aligned} u &= \frac{\sqrt{2}}{2}\left(\sqrt{2}\cos(t)\right) - \frac{\sqrt{2}}{2}\left(-\sqrt{2} + \sqrt{2}\sin(t)\right) \\ &= \cos(t) - \sin(t) + 1. \end{aligned}$$

Similarly,

$$\begin{aligned} v &= \frac{\sqrt{2}}{2}\left(\sqrt{2}\cos(t)\right) + \frac{\sqrt{2}}{2}\left(-\sqrt{2} + \sqrt{2}\sin(t)\right) \\ &= \cos(t) + \sin(t) - 1. \end{aligned}$$

Note that

$$(u - 1)^2 + (v + 1)^2 = 2.$$

That is, the circle of radius $\sqrt{2}$ centered at the point $(0, -\sqrt{2})$ in the xy-plane corresponds to the circle of radius $\sqrt{2}$ centered at the point $(1, -1)$ in the uv-plane. This region can be described by $0 \le u \le \sqrt{2}$ and

$$-1 + \sqrt{2 - (u-1)^2} \le v \le 2 - \sqrt{2 - u^2}.$$

Finally, we have the Jacobian and its determinant given below.

#1: $\text{JACOBIAN}\left(\left[\dfrac{\sqrt{2}}{2} \cdot u + \dfrac{\sqrt{2}}{2} \cdot v, \; -\dfrac{\sqrt{2}}{2} \cdot u + \dfrac{\sqrt{2}}{2} \cdot v\right], \; [u, v]\right)$

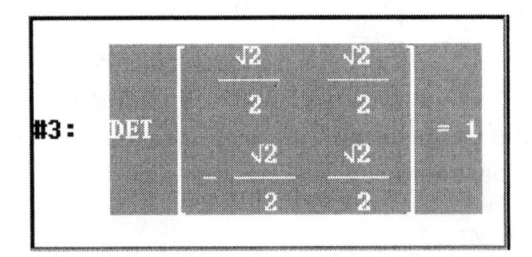

#2: $\begin{bmatrix} \dfrac{\sqrt{2}}{2} & \dfrac{\sqrt{2}}{2} \\ -\dfrac{\sqrt{2}}{2} & \dfrac{\sqrt{2}}{2} \end{bmatrix}$

#3: $\text{DET}\begin{bmatrix} \dfrac{\sqrt{2}}{2} & \dfrac{\sqrt{2}}{2} \\ -\dfrac{\sqrt{2}}{2} & \dfrac{\sqrt{2}}{2} \end{bmatrix} = 1$

(Don't forget to **Load** the utility file **Vector.mth** prior to using the **JACOBIAN** command.) Therefore,

$$\begin{aligned}\iint_R xy\,dx\,dy &= \int_0^{\sqrt{2}} \int_{-1+\sqrt{2-(u-1)^2}}^{2-\sqrt{2-u^2}} \left(\dfrac{\sqrt{2}}{2}u + \dfrac{\sqrt{2}}{2}v\right)\left(-\dfrac{\sqrt{2}}{2}u + \dfrac{\sqrt{2}}{2}v\right) dv\,du \\ &= 2\sqrt{2} + \dfrac{7}{6} - \pi - \dfrac{1}{2}\sqrt{\left(-1 + 2\sqrt{2}\right)} - \dfrac{1}{6}\sqrt{\left(-1 + 2\sqrt{2}\right)}\sqrt{2} \\ &= -.14131 \end{aligned}$$

(as **DERIVE** will show).

The situation for triple integrals is more complicated from a geometric point of view. Algebraically, it is essentially the same. If T is a C^1 map

6.6. CHANGE OF VARIABLES

which maps a region S is in uvw-space one-to-one and onto a region R in xyz-space and we make the change of variables

$$(x, y, z) = T(u, v, w) = (x(u, v, w), y(u, v, w), z(u, v, w))$$

then the Jacobian derivative is given by

$$J(u, v, w) = \begin{pmatrix} x_u & x_v & x_w \\ y_u & y_v & y_w \\ z_u & z_v & z_w \end{pmatrix}$$

and the Jacobian determinant is given by

$$\frac{\partial(x, y, z)}{\partial(u, v, w)} = \det(J(u, v, w)).$$

Furthermore,

$$\iiint_R f(x,,z) \, dxdydz = \iiint_S f(J(u, v, w)) \left| \frac{\partial(x, y, z)}{\partial(u, v, w)} \right| dudvdw.$$

The Jacobian determinant can be computed using derive in the same manner as shown above in the case of two variables. We simply need to **Author** the **Expression**

JACOBIAN([x(u, v, w), y(u, v, w), z(u, v, w)], [u, v, w]),

click on the ■ icon, and then apply the **DET** command to obtain the result.

Exercises:

1. Plot the image (in the xy-plane) of both the unit square and the unit disk (in the uv-plane) under each of the following transformations. Also compute the Jacobian determinant of the transformation and describe the set of points in the uv-plane on which it is zero.

    (a) $(x, y) = (u + v, v^2 - u^2)$.
    (b) $(x, y) = (u + v, uv)$.
    (c) $(x, y) = (u^2 - v^2, u^2 + v^2)$.

(d) $(x, y) = (u^2 - v^2, 2uv)$.

2. Let R be the trapezoid in the xy-plane bounded by the lines $y = 1 - x$, $y = 2 - x$ $y = -3x$ and $y = x/2$. Sketch this region. Then simplify the computation of the integral over R of
$$f(x, y) = xy$$
with a change of variables that rotates R through a clockwise angle of $\pi/4$ radians.

3. Let R be the triangle in the xy-plane with vertices $(2, -7)$, $(6, 13)$ and $(3, 16)$. Compute the integral of
$$f(x, y) = xy$$
over R by using a change of variables which maps the triangle in the uv-plane with vertices $(0, 0)$, $(1, 0)$ and $(0, 1)$ onto R.

4. Let R be the region in the xy-plane bounded by $y = \sqrt{x}$, $y = 0$ and $y = 2\sqrt{x} - 1$. Compute the integral of
$$f(x, y) = \cos(\sqrt{x} - y)$$
over R by means of the change of variables
$$u = \sqrt{x} \text{ and } v = \sqrt{x} - y.$$
(Note: You will need to solve for x and y.)

5. Let R be the rectangle in the xy-plane bounded by the lines $x - y = 0$, $x + y = 0$, $x - y = \pi/2$, and $x + y = \pi$. Compute the integral of
$$f(x, y) = 2x - 3y$$
over R by making a change of variables which maps the square with vertices $(0, 0)$, $(1, 0)$, $(1, 1)$ and $(0, 1)$ in the uv-plane onto the rectangle above.

6. Compute the area of the region bounded by the curves
$$x + xy - y^2 = 1, \ x + xy - y^2 = 3, \ -x + y = -2 \text{ and } -x + y = 0$$
by making the change of variables
$$(u, v) = (x + xy - y^2, -x + y).$$

6.6. CHANGE OF VARIABLES

7. Compute the Jacobian determinants for each of the following change of variables.

   (a) $(x, y, z) = (u^2 - v + w, 2uv - v^2 + w, 3u + 2v - w^2)$.
   (b) $(x, y) = \left(\frac{1}{3} r \cos(\theta), \frac{1}{4} r \sin(\theta)\right)$
   (c) $(x, y, z) = (2\rho \cos(\theta) \sin(\phi), 3\rho \sin(\theta) \sin(\phi), \rho \cos(\phi))$.

8. Let R be the three dimensional region inside of the ellipsoid described by
$$x^2 + \frac{1}{4}y^2 + \frac{1}{9}z^2 = 1.$$
Compute the integral over R of
$$f(x, y, z) = x + y + z$$
by using the change of variables $u = x$, $v = \frac{1}{2}y$ and $w = \frac{1}{3}z$, and then converting to spherical coordinates.

Chapter 7

Vector Calculus

Vector calculus is the study of vector fields and centers around three higher-dimensional versions of the Fundamental Theorem of Calculus, known as Green's Theorem, Stokes' Theorem and the Divergence Theorem.

- Vector Calculus is the subject of Chapter 13 of Stewart's **Multivariable Calculus: Concepts and Contexts**.

7.1 Vector Fields

- Vector fields are defined and discussed in Section 13.1 of Stewart's text.

The **DERIVE** has a built in command **DIRECTION_FIELD**, in the utility file **Ode_appr.mth**, for sketching two dimensional vector fields. Unfortunately, this routine only sketches line segments without arrows. As a result, the user can not use this routine to determine the direction of the vectors in the direction field. Also, in our setting the arguments might seem a little unnatural. For example, if the vector field has the form

$$\mathbf{F}(x, y) = \langle p(x, y), q(x, y) \rangle$$

and we want to sketch the vector field for $a \leq x \leq b$ and $c \leq y \leq d$, then we must author the expression

DIRECTION_FIELD(q(x,y)/p(x,y), x, a, b, m, y, c, d, n),

click on the ▪ icon, and plot the result in a **2-D Plot** window. The values m and n refer to the number of arrows in each direction. We demonstrate the use of this command below.

Example 1 *Plot the vector field* $\mathbf{F}(x,y) = \langle x - y, x \rangle$ *in the rectangle* $-2 \leq x \leq 2$, $-1 \leq y \leq 1$.

Start by **Loading** the utility file **Ode_appr.mth**. Then the **DIRECTION_FIELD** command can be used to sketch the vector field. Simply author the expression

$$\text{DIRECTION\_FIELD}(x/x - y, x, -2, 2, 10, y, -1, 1, 5),$$

click on the [=] icon, and plot the result in a **2-D Plot** window.

Notice that we have used ten (10) vectors in the x direction and five (5) vectors in the y direction.

One of the important questions associated with a given vector field $\mathbf{F}(x, y)$ is whether it is a conservative vector field. Recall from Section 13.1 in Stewart's text that a vector field

$$\mathbf{F}(x, y) = \langle p(x, y), q(x, y) \rangle$$

defined on an open, connected region D is said to be a conservative vector field (or gradient field) if and only if there is a potential function f so that

$$\nabla f(x, y) = \mathbf{F}(x, y)$$

for all $(x, y) \in D$. Stewart shows that this is true if and only if

$$\frac{\partial p(x, y)}{\partial y} = \frac{\partial q(x, y)}{\partial x} \tag{7.1}$$

for all $(x, y) \in D$. When this is the case, we can integrate to obtain a potential function f using the formula

$$f(x, y) = \int_a^x p(t, y) dt + \int_b^y q(a, t) dt \tag{7.2}$$

7.1. VECTOR FIELDS

where (a, b) is any point in D. We give an example below to show how the related computations can be performed using **DERIVE**.

Example 2 *Show that the vector field*

$$\mathbf{F}(x, y) = \langle 1 + 2xy - y\sin(xy), x^2 - x\sin(xy) \rangle$$

is conservative and find an associated potential function f.

We start by defining the components of $\mathbf{F}$ as p and q.

#1: $P(x, y) := 1 + 2 \cdot x \cdot y - y \cdot \text{SIN}(x \cdot y)$

#2: $Q(x, y) := x^2 - x \cdot \text{SIN}(x \cdot y)$

Then we compute $q_x(x, y) - p_y(x, y)$, and verify that it is zero.

#3: $\dfrac{d}{dx} Q(x, y) - \dfrac{d}{dy} P(x, y)$

#4: 0

Since the equation (7.1) is satisfied, there must be a function f so that $\nabla f(x, y) = \mathbf{F}(x, y)$. We use the formula (7.2) below with the values $a = b = 0$ to find such an f.

#5: $\displaystyle\int_0^x P(t, y)\, dt + \int_0^y Q(0, t)\, dt$

#6: $\text{COS}(x \cdot y) + x^2 \cdot y + x - 1$

Determining whether a 3-dimensional vector field is a conservative vector field (a gradient field) is only slightly more complicated. If

$$\mathbf{F}(x, y, z) = \langle p(x, y), q(x, y), r(x, y) \rangle$$

is a vector field defined on an open, connected region E in R^3 then $\mathbf{F}$ is a conservative vector field if and only if

$$p_y = q_x, \quad p_z = r_x, \quad q_z = r_y \tag{7.3}$$

on E. In this case a potential function f satisfying

$$\nabla f(x,y,z) = \mathbf{F}(x,y,z)$$

is given by

$$f(x,y,z) = \int_a^x p(t,y,z)dt + \int_b^y q(a,t,z)dt + \int_c^z r(a,b,t)dt \tag{7.4}$$

for any point $(a,b,c) \in E$.

Example 3 *Show that*

$$f(x,y,z) = \langle 2xz - y\sin(x), -2z + \cos(x), x^2 - 2y \rangle$$

is a conservative vector field.

We define the functions p, q and r below.

```
#1:  P(x, y, z) := 2·x·z - y·SIN(x)
#2:  Q(x, y, z) := - 2·z + COS(x)
                         2
#3:  R(x, y, z) := x  - 2·y
```

Then we verify that the conditions in (7.3) hold by creating the vector

$$[p_y - q_x, p_z - r_x, q_z - r_y]$$

and verifying that it is $[0,0,0,]$.

```
#4:  [d/dy P(x, y, z) - d/dx Q(x, y, z), d/dz P(x, y, z) - d/d
#5:  [0, 0, 0]
```

7.1. VECTOR FIELDS

Since the conditions in (7.3) hold, we can compute a potential for f by using (7.4).

$$\#6: \quad \int_0^x P(t, y, z)\, dt + \int_0^y Q(0, t, z)\, dt + \int_0^z R(0, 0, t)\, dt$$

$$\#7: \quad y\cos(x) + x^2 z - 2yz$$

Consequently, a potential function is given by

$$f(x, y, z) = y\cos(x) + x^2 z - 2yz.$$

Notice that we can easily verify that the function f shown is a potential function for our vector field by computing the partial derivatives

$$f_x(x, y, z), \ f_y(x, y, z) \text{ and } f_z(x, y, z)$$

showing that

$$f_x = p, \ f_y = q, \ f_z = r.$$

This can also be accomplished by using **DERIVE**'s built in gradient command **GRAD**.

Exercises:

1. Define
$$\mathbf{F}(x, y) = \langle x + y, x - y \rangle.$$

    (a) Plot the vector field given by $\mathbf{F}(x, y)$.

    (b) Determine whether $\mathbf{F}$ is a conservative vector field.

    (c) If $\mathbf{F}$ is a conservative vector field then find a potential function for $\mathbf{F}$.

2. Repeat problem 1 for the vector field
$$\mathbf{F}(x, y) = \langle x - y, x + y \rangle.$$

3. Repeat problem 1 for the vector field
$$\mathbf{F}(x,y) = \langle \cos(x+y) - x\sin(x+y), -x\sin(x+y) \rangle.$$

4. Repeat problem 1 for the vector field
$$\mathbf{F}(x,y) = \left\langle \frac{-y}{x^2+y^2}, \frac{x}{x^2+y^2} \right\rangle \text{ on } D = \{(x,y) \mid y > 0\}.$$

5. Repeat problem 1 for the vector field
$$\mathbf{F}(x,y) = \langle x - xy, -y + xy \rangle.$$

6. Define
$$\mathbf{F}(x,y,z) = \langle x+y-z, x-y, z-x \rangle.$$

   (a) Determine whether $\mathbf{F}$ is a conservative vector field.

   (b) If $\mathbf{F}$ is a conservative vector field then find a potential function for $\mathbf{F}$.

7. Repeat problem 6 for the vector field
$$\mathbf{F}(x,y,z) = \left\langle -2\cos xz + 2x(\sin xz)z - z^3, 2z, 2y + 2x^2 \sin xz - 3xz^2 \right\rangle.$$

8. Repeat problem 6 for the vector field
$$\mathbf{F}(x,y,z) = \frac{\langle x,y,z \rangle}{\sqrt{x^2+y^2+z^2}} \text{ on } D = \{(x,y,z) \mid x,y,z \neq 0\}.$$

7.2 Line Integrals

- Line integrals are the subject of Sections 13.2 and 13.3 of Stewart's **Multivariable Calculus: Concepts and Contexts**.

Line integrals of scalar and vector functions are easy to compute in **DERIVE**. The examples below demonstrate the basic ideas.

7.2. LINE INTEGRALS

Example 4 *Compute the line integral*
$$\int_C f \, ds$$
where f is given by
$$f(x,y) = \frac{xy}{1+x+2y}$$
and C is the unit quarter circle in the first quadrant from $(1,0)$ to $(0,1)$.

We start by parameterizing the quarter circle as
$$\mathbf{r}(t) = \langle \cos(t), \sin(t) \rangle \text{ for } 0 \le t \le \frac{\pi}{2}.$$
Then
$$ds = \|\mathbf{v}(t)\| \, dt,$$
where $\mathbf{v}(t) = \mathbf{r}'(t)$, and the line integral is given by
$$\int_C f \, ds = \int_0^{\pi/2} f(\mathbf{r}(t)) \, |\mathbf{r}'(t)| \, dt.$$
We define f and compute $|\mathbf{r}'(t)|$ below.

```
#1:   F(x, y) := ───────────
                  x·y
                 1 + x + 2·y

#2:   R(t) := [COS(t), SIN(t)]

#3:   |R'(t)| = 1
```

Then we calculate the line integral.

```
        π/2
#4:     ∫    F(COS(t), SIN(t)) dt
        0

#5:   0.168183
```

Line integrals along curves in 3-dimensional space are also easy to compute.

Example 5 *Compute the line integral*

$$\int_C f \, ds$$

where

$$f(x, y, z) = \frac{1}{x^2 + y^2 + z^2}$$

and C is the portion of the helix parameterized by

$$r(t) = \langle \cos(2\pi t), \sin(2\pi t), t/2 \rangle \text{ for } 0 \leq t \leq 2.$$

The solution process is no different from the example above. The screens below show the process.

```
#1:  F(x, y, z) := 1/(x² + y² + z²)

#2:  R(t) := [COS(2·π·t), SIN(2·π·t), t/2]

#3:  |R'(t)| = √(16·π² + 1)/2
```

```
          2
         ⌠  F[COS(2·π·t), SIN(28·π·t), t/2]·√(16·π² + 1)
#4:      ⎮  ─────────────────────────────────────────── dt
         ⌡                      2
         0

#5:  14.7959
```

Line integrals of vector fields are also easy to compute with **DERIVE**.

Example 6 *Compute the line integral*

$$\int_C F \cdot dr$$

where

$$F(x, y) = \langle -y, x \rangle$$

and C is the portion of the curve $y = x^4$ *from* $(0, 0)$ *to* $(1, 1)$.

7.3. GREEN'S THEOREM

Note that C can be parameterized by
$$r(t) = \langle t, t^4 \rangle \text{ for } 0 \leq t \leq 1.$$

Also, recall that
$$\mathbf{dr} = r'(t)\, dt = v(t)\, dt.$$

The computations are shown below.

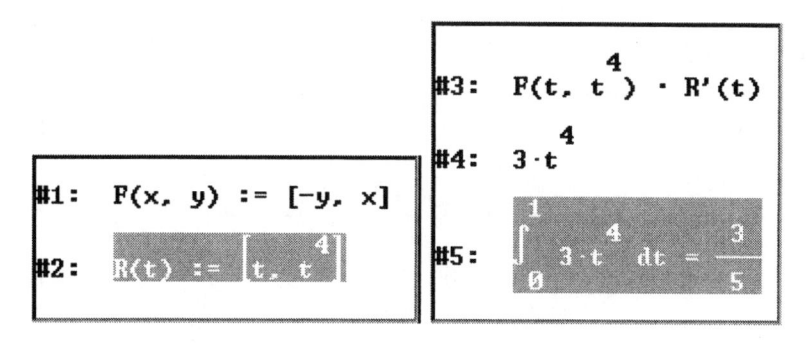

```
#1:  F(x, y) := [-y, x]
#2:  R(t) := [t, t^4]
#3:  F(t, t^4) · R'(t)
#4:  3·t^4
#5:  ∫₀¹ 3·t^4 dt = 3/5
```

7.3 Green's Theorem

- Section 13.4 of Stewart's **Multivariable Calculus: Concepts and Contexts** states and proves a special case of Green's Theorem, and defines the notion of a positively oriented curve.

Recall that Green's Theorem states that if D is a region in the plane whose boundary ∂D consists of one or more positively-oriented, piecewise-smooth, simple closed curves, and
$$\mathbf{F}(x, y) = \langle p(x, y), q(x, y) \rangle$$
is a continuously differentiable vector field on an open region containing $D \cup \partial D$ then
$$\iint_D \left(\frac{\partial q}{\partial x} - \frac{\partial p}{\partial y} \right) dA = \int_{\partial D} \mathbf{F} \cdot \mathbf{dr}. \tag{7.5}$$

Example 7 *Verify Green's theorem if the vector field is*
$$\mathbf{F}(x, y) = \langle xy, x - y \rangle$$
and D is the open unit disk.

We define p and q below and create the integrand.

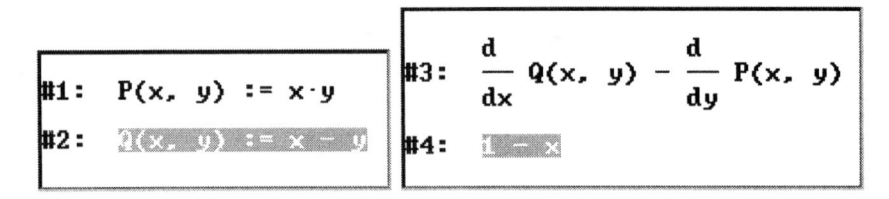

#1: $P(x, y) := x \cdot y$

#2: $Q(x, y) := x - y$

#3: $\dfrac{d}{dx} Q(x, y) - \dfrac{d}{dy} P(x, y)$

#4: $1 - x$

Then we compute the double integral on the left hand side of (7.5) by using polar coordinates.

#5: $G(x, y) := 1 - x$

#6: $\displaystyle\int_0^{2\pi} \int_0^1 G(r \cdot \cos(\theta), r \cdot \sin(\theta)) \cdot r \, dr \, d\theta = \pi$

Now we define **F**, parameterize ∂D and compute the line integral on the right hand side of (7.5).

#7: $F(x, y) := [x \cdot y, x - y]$

#8: $R(t) := [\cos(t), \sin(t)]$

#9: $F(\cos(t), \sin(t)) \cdot R'(t)$

#10: $\displaystyle\int_0^{2\pi} F(\cos(t), \sin(t)) \cdot R'(t) \, dt = \pi$

Clearly, the results are the same!

Exercises:

1. A circular loop of wire with radius 1 has mass density given by the square of the distance from one fixed point on the loop. Find the mass of the wire and its center of mass. (Hint: Think of the wire as the unit circle and assume that the fixed point from which distance is measured is $(-1, 0)$.)

2. A wire of uniform density is bent in the shape of the graph of $y = x^2$ for $-1 \leq x \leq 1$. Find its center of mass.

3. Find the work done on a particle that moves counter clockwise along the unit quarter-circle in the first quadrant through a force field given by
$$\mathbf{F}(x,y) = \langle x - xy, -y + xy \rangle.$$

4. A particle travels around the unit square ($0 \leq x \leq 1$, $0 \leq y \leq 1$) exactly once, moving counter clockwise along the edges of the square in the presence of a force field given by
$$\mathbf{F}(x,y) = \langle xy, x - y \rangle.$$
Find the work done on the particle by the force field. First do this directly by computing a line integral. Then redo the calculation with the help of Green's Theorem.

5. A particle travels once around the circle of radius ρ centered at the origin in a counterclockwise direction, while acted upon by a force field given by
$$\mathbf{F}(x,y) = \langle x^2 - y^3 + 6y, x^3 - y^2 - 3x \rangle.$$
Is there a value of ρ such that the work done on the particle is zero?

6. A particle moves counter clockwise around the edge of the region bounded by $y = 1 - x^2$ and $y = x^2 - 1$ while being acted upon by a force field given by
$$\mathbf{F}(x,y) = \langle x^2 - y^3, x^3 + y^2 \rangle.$$
Find the work done on the particle during one trip around the path. First do this directly by computing a line integral. Then redo the calculation with the help of Green's Theorem.

7.4 Surface Integrals

- See Section 13.6 of Stewart's **Multivariable Calculus: Concepts and Contexts** for information related to surface integrals.

According to Stewart, if S is a surface parameterized by a continuously differentiable function $\mathbf{r}(u,v)$ for $(u,v) \in D$, and f is a continuous function defined on S, then the surface integral of f over S is given by

$$\iint_S f(x,y,z) \, d\mathbf{S} = \iint_D f(\mathbf{r}(u,v)) \, |\mathbf{r}_u \times \mathbf{r}_v| \, dA. \qquad (7.6)$$

In the case when S is the graph of $z = g(x,y)$ this formula becomes

$$\int\int_S f(x,y,z)dS = \int\int_D f(x,y,g(x,y))\left((g_x)^2 + (g_y)^2 + 1\right)^{1/2} dA. \quad (7.7)$$

Surface integrals are easy to compute with **DERIVE**.

Example 8 *Compute the surface integral of*

$$f(x,y,z) = \frac{1}{1+x^2+y^2}$$

over the top half of the unit sphere.

The top half of the unit sphere can be parameterized with spherical coordinates by

$$\mathbf{r}(\phi, \theta) = \langle \cos(\theta)\sin(\phi), \sin(\theta)\sin(\phi), \cos(\phi) \rangle$$

for $0 \leq \theta \leq 2\pi$ and $0 \leq \phi \leq \pi/2$. The formula given in (7.6) is used below to compute the surface integral. Notice that we have used u and v instead of ϕ and θ. The screen below shows the definition of f and $\mathbf{r}$, and the formation of the partial derivatives of $\mathbf{r}$ and $|\mathbf{r}_u(u,v) \times \mathbf{r}_v(u,v)|$.

```
#1:   F(x, y, z) := ─────────────
                      1 + x² + y²

#2:   R(φ, θ) := [COS(θ)·SIN(φ), SIN(θ)·SIN(φ), COS(φ)]

#3:   ‖CROSS(d/dφ R(φ,θ), d/dθ R(φ,θ))‖ = |SIN(φ)|
```

Now, note that the restriction $0 \leq v \leq \pi/2$ implies that $|\sin(v)| = \sin(v)$. Consequently, we can compute the surface integral as shown below.

7.4. SURFACE INTEGRALS

```
#4:     F(COS(θ)·SIN(φ), SIN(θ)·SIN(φ), COS(φ))·SIN(φ)

                SIN(φ)
                ──────
#5:                2
            SIN(φ)  + 1

         2·π   π/2
          ⌠     ⌠      SIN(φ)
#6:       ⎮     ⎮   ──────────── dφ dθ  =  √2·π·LN(√2 + 1)
          ⎮     ⎮         2
          ⌡     ⌡    SIN(φ)  + 1
          0     0
```

Example 9 *Compute the flux of the vector field*

$$\mathbf{F}(x,y,z) = \langle -y, z, x \rangle$$

across the surface given by the portion of the paraboloid

$$z = 1 - x^2 - y^2$$

which lies above the unit disk.

In this case, if we define

$$g(x,y) = 1 - x^2 - y^2$$

then we need to compute the surface integral of

$$\mathbf{F}(x, y, g(x,y)) \cdot \langle -g_x, -g_y, 1 \rangle .$$

The screen below shows the generation of the functions **F** and *g* above.

```
#1:    F(x, y, z) := [-y, z, x]

                              2    2
#2:    G(x, y) := 1 - x  - y
```

Now we create the expression $\mathbf{F}(x, y, g(x,y)) \cdot \langle -g_x, -g_y, 1 \rangle$.

```
#3:   F(x, y, G(x, y)) · [- d/dx G(x, y), - d/dy G(x, y), 1]

#4:   - 2·x²·y + x·(1 - 2·y) - 2·y·(y² - 1)
```

Finally, we use (7.7) to compute the flux. The first screen below shows the formation of the integrand

$$\mathbf{F}(x, y, g(x,y)) \cdot \langle -g_x, -g_y, 1 \rangle \left((g_x)^2 + (g_y)^2 + 1 \right)^{1/2}$$

and some simplification. The output is too long to include in the screen shot.

```
#5:   (- 2·x²·y + x·(1 - 2·y) - 2·y·(y²

#6:   - √(4·x² + 4·y² + 1)·(2·x²·y + x·(

#7:   H(x, y) := - √(4·x² + 4·y² + 1)·(2

           2·π   1
#8:        ∫    ∫  H(r·COS(θ), r·SIN(θ))·r dr dθ
           0    0

#9:   ▯
```

You should ask yourself why the symmetry in this problem causes the flux to be zero.

Exercises: Note that some of the integrations below must be done numerically.

1. Find the area of the portion of the surface

$$z = x^2 + e^y$$

that lies above:

7.4. SURFACE INTEGRALS

(a) the disk $x^2 + y^2 \leq 1$;

(b) the square given by $-1 \leq x \leq 1$, $-1 \leq y \leq 1$.

2. A thin spherical shell, with radius 1, has mass density given by the square of the distance from one fixed point on the shell. Find the mass of the shell. (Hint: Think of the shell as the unit sphere and assume that the fixed point from which distance is measured is $(0, 0, -1)$.)

3. A thin metal sheet is bent in the shape of the portion of the surface $z = x^2 + y^2$ that lies above the square given by $-1 \leq x \leq 1$, $-1 \leq y \leq 1$. Its mass density is $\rho(x, y, z) = z$. Find its mass.

4. A thin metal sheet is bent in the shape of the portion of the surface $z = x^2 + y^2$ that lies above the disk given by $x^2 + y^2 \leq 1$. Its mass density is $\rho(x, y, z) = z$. Find its mass.

5. Let $f(x, y, z) = xyz$. Find the flux of ∇f across the portion of the plane

$$2x + 3y + z = 6$$

on which $x \geq 0$, $y \geq 0$, $z \geq 0$.

6. Let $\mathbf{F}(x, y, z) = \langle -y, z, x \rangle$ and let S be the portion of the surface $z = x^2 + y^2$ that lies above the unit quarter disk in the first quadrant of the xy-plane. Compute the flux of $\mathbf{F}$ across S.

7. Let $\mathbf{F}(x, y, z) = \langle x, y, z \rangle$ and let S be the sphere

$$x^2 + y^2 + (z - 1)^2 = 1.$$

Compute the flux of $\mathbf{F}$ across S.

8. Let $\mathbf{F}(x, y, z) = \langle x, y, z \rangle$ and let S be the cylindrical surface given by

$$x^2 + y^2 = 1 \text{ and } 0 \leq z \leq 1.$$

Compute the flux of $\mathbf{F}$ across S.

7.5 Stokes' Theorem

- Stokes' Theorem is the subject of Section 13.7 of Stewart's **Multivariable Calculus: Concepts and Contexts**.

You should recall from Stewart's text that if S is an oriented, piecewise smooth surface whose boundary (or edge) is a simple, closed, piecewise smooth, positively oriented curve C, and $\mathbf{F}$ is a continuously differentiable vector field on an open region containing S then

$$\int_C \mathbf{F} \cdot \mathbf{dr} = \int\int_S \text{curl}(\mathbf{F}) \cdot \mathbf{dS}. \tag{7.8}$$

where $\mathbf{F} = \langle p, q, r \rangle$ implies

$$\text{curl}(\mathbf{F}) = \nabla \times \mathbf{F} = \langle r_y - q_z, p_z - r_x, q_x - p_y \rangle.$$

Example 10 *Verify the conclusion of Stokes' Theorem if S is the portion of the paraboloid*

$$z = x^2 + y^2$$

that lies under the plane $z = 1$, and

$$\mathbf{F}(x, y, z) = \langle y - z^2, z - x^2, x - y^2 \rangle.$$

We begin by computing the line integral in (7.8). Note that in this case C is parameterized by

$$\langle \cos(t), \sin(t), 1 \rangle \text{ for } 0 \leq t \leq 2\pi.$$

```
#1:   F(x, y, z) := [y - z², z - x², x - y²]
#2:   R(t) := [COS(t), SIN(t), 1]
           2·π
#3:    ∫    F(COS(t), SIN(t), 1) · R'(t) dt = -π
           0
```

7.6. THE DIVERGENCE THEOREM

Now let's compute the surface integral in (7.8). Notice that in this case S is the portion of the paraboloid $z = x^2 + y^2$ that lies above the unit disk in the xy-plane. Consequently, the surface integral in (7.8) is given by

$$\iint_D \text{curl}(\mathbf{F}(x, y, g(x, y))) \cdot \langle -g_x, -g_y, 1 \rangle \, dA$$

where $g(x, y) = x^2 + y^2$ and D is the unit disk in the xy-plane. We start by defining the function g.

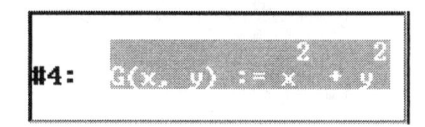

#4: $G(x, y) := x^2 + y^2$

Then we create the integrand for the surface integral by using the command **CURL**.

#5: $\text{CURL}(F(x, y, G(x, y))) \cdot \left[-\dfrac{d}{dx} G(x, y), -\dfrac{d}{dy} G(x, y), 1 \right]$

#6: $4 \cdot x^2 \cdot y + 4 \cdot x \cdot y + 4 \cdot y^3 + 2 \cdot y - 1$

#7: $H(x, y) := 4 \cdot x^2 \cdot y + 4 \cdot x \cdot y + 4 \cdot y^3 + 2 \cdot y - 1$

Finally, the integral is computed below using polar coordinates.

#8: $\displaystyle\int_0^{2\pi} \int_0^1 H(s \cdot \text{COS}(\theta), s \cdot \text{SIN}(\theta)) \cdot s \, ds \, d\theta = -\pi$

As above, we get the value $-\pi$, verifying Stokes' theorem.

7.6 The Divergence Theorem

- Section 13.8 of Stewart's **Multivariable Calculus: Concepts and Contexts** states and proves a special case of the Divergence Theorem.

Recall from Stewart's text that if E is a solid region whose boundary ∂E is a piecewise smooth surface, and the vector field $\mathbf{F}$ is continuously differentiable in a region containing E, then

$$\iint_{\partial E} \mathbf{F} \cdot d\mathbf{S} = \iiint_E \operatorname{div}(\mathbf{F}) \, dV, \tag{7.9}$$

where $\operatorname{div}(\mathbf{F})$ denotes the divergence of $\mathbf{F}$. Note that this equation states that the flux of $\mathbf{F}$ across ∂E is the integral of the divergence of $\mathbf{F}$ over E.

Example 11 *Use the Divergence Theorem to compute the flux of the vector field*

$$\mathbf{F}(x, y, z) = \left\langle x^2 + \cos(y^2 z), x - yz + z, xyz \right\rangle$$

across the surface of the box E described by $0 \leq x \leq 1, 0 \leq y \leq 1, 0 \leq z \leq 1$.

According to (7.9) we can compute the flux by calculating

$$\iiint_E \operatorname{div}(\mathbf{F}) \, dV.$$

The calculations are shown in the screen below. Notice that the command **DIV** computes the divergence.

```
#1:  F(x, y, z) := [x^2 + COS(y^2 ·z), x - y·z + z, x·y·z]

              1  1  1
#2:           ∫  ∫  ∫  DIV(F(x, y, z)) dx dy dz  =  3/4
              0  0  0
```

Exercises:

1. Use Stokes' Theorem to evaluate the line integral of

$$\mathbf{F}(x, y, z) = \left\langle yz - \cos(\pi x), z - \sin(\pi y), xy + z^3 \right\rangle$$

along the boundary of the parallelogram that lies in the plane $x + y + z = 2$ and above the unit square in the xy-plane, oriented so that it corresponds to a counterclockwise orientation of the edges of the unit square in the xy-plane.

7.6. THE DIVERGENCE THEOREM

2. Use Stokes' Theorem to evaluate the line integral
$$\int_C \mathbf{F} \cdot d\mathbf{r}$$
where
$$\mathbf{F}(x,y,z) = \langle z - \cos(x), x^2 - z^2, x^2 + y^2 + z^2 \rangle$$
and C is the intersection of the surface $z = x^2 + y^3$ and the cylinder $x^2 + y^2 = 1$, oriented in a way that corresponds to a counterclockwise orientation of its projection onto the xy-plane.

3. Use Stokes' Theorem to evaluate the line integral
$$\int_C \mathbf{F} \cdot d\mathbf{r}$$
where
$$\mathbf{F}(x,y,z) = \langle z + x^5, x + y^{2/3}, xz + \cos(\pi y) \rangle$$
and C is the intersection of the surface $z = 1 - x^2 y$ and the surface of the box given by $-1 \leq x \leq 1$, $-1 \leq y \leq 1$, $0 \leq z \leq 2$, oriented in a way that corresponds to a counterclockwise orientation of tis projection onto the xy-plane.

4. Verify the statement of the Divergence Theorem where
$$\mathbf{F}(x,y,z) = \langle xy, y^2, z + xy \rangle$$
and E is the cylinder given by $x^2 + y^2 \leq 1$ and $0 \leq z \leq 1$. (Note: There are three parts to the surface ∂E.)

5. Use the Divergence Theorem to compute the flux of
$$\mathbf{F}(x,y,z) = \langle x^2 - yz, y^2 + xz, z^2 - xy \rangle$$
across the surface of the cylinder given by $x^2 + y^2 \leq 1$ and $0 \leq z \leq 1$.

6. Let E be the hemisphere described by $x^2 + y^2 + z^2 \leq 1$ and $0 \leq z \leq 1$. By the Divergence Theorem, the flux of
$$\mathbf{F}(x,y,z) = \langle x^2 z + \cos(\pi z), y - 2xyz, -z + x + y \rangle$$
across ∂E is zero, since $\text{div}(\mathbf{F}) = 0$. Make use of this to help compute the flux of $\mathbf{F}$ across the spherical (top) part of the surface ∂E.

7. Suppose that a function f satisfies
$$\nabla^2 f(x,y,z) = e^{-(x^2+y^2+z^2)}$$
in the interior of the unit sphere, where $\nabla^2 f(x,y,z)$ denotes the Laplacian of f. Find the flux of ∇f across the surface of the unit sphere.

Chapter 8

Projects

8.1 Osculating Circles

- Background material is found in Sections 10.1, 10.3 and 10.4 of Stewart's **Multivariable Calculus: Concepts and Contexts**.

Let C be a curve in the plane, parametrized by $\mathbf{r}(t) = \langle x(t), y(t) \rangle$. Recall that the unit tangent vector at $(x(t), y(t))$ is given by

$$\mathbf{T}(t) = \frac{1}{|\mathbf{r}'(t)|} \mathbf{r}'(t)$$

and the curvature at $\mathbf{r}(t)$ is given by

$$\kappa(t) = \frac{|\mathbf{T}'(t)|}{|\mathbf{r}'(t)|}.$$

One interpretation of curvature is that its reciprocal is the radius of the circle that best approximates the curve at the given point. However, since $\kappa(t)$ is by definition a nonnegative quantity, it does not indicate on which side of the curve this best approximating circle should lie; i.e., it only indicates the *amount* of "bend" in the curve and does not indicate in which *direction* the curve is bending. So it will be useful to define a **signed curvature** $\kappa_\pm(t)$ by

$$\kappa_\pm(t) = \kappa(t) \operatorname{sign}\left(x'(t) y''(t) - y'(t) x''(t)\right).$$

1. Explain carefully why the signed curvature will be positive when the curve bends to the left and negative when it bends to the right. (*Hint*: Think about the cross product of $\langle x'(t), y'(t), 0 \rangle$ and $\langle x''(t), y''(t), 0 \rangle$.)

The **osculating circle** to the curve at a given point is the circle that best approximates the curve near that point. It is tangent to the curve, has the same curvature, and lies on the side of the curve toward which the curve is bending. An excellent picture is given in Figure 9 of Stewart's Section 10.3.

2. Let
$$\mathbf{c}(t) = \mathbf{r}(t) + \frac{1}{\kappa_{\pm}(t)} \mathbf{n}(t),$$
where
$$\mathbf{n}(t) = \frac{1}{|\mathbf{r}'(t)|} \langle -y'(t), x'(t) \rangle.$$

(a) Show that $\mathbf{n}(t)$ is a unit vector that is orthogonal to the curve at $(x(t), y(t))$.

(b) Show that $\mathbf{n}(t)$ points to the left of the curve at $(x(t), y(t))$.

(c) Show that $\mathbf{c}(t)$ is the position vector of the center of the osculating circle to the curve at $(x(t), y(t))$.

3. Create a plot that shows the curve $y = \sin(x)$, parametrized by
$$\mathbf{r}(t) = \langle t, \sin(t) \rangle$$
along with osculating circles at $(\pi/2, 1)$ and $(3\pi/2, -1)$.

4. Create a plot that shows the curve $\mathbf{r}(t) = \langle \sin(t), \sin(2t) \rangle$ along with osculating circles at points corresponding to $t = \pi/5$, $\pi/4$, and $\pi/3$.

5. Now recall that the principle unit normal vector is given by
$$\mathbf{N}(t) = \frac{1}{|\mathbf{a}(t) - proj_{\mathbf{v}(t)}\mathbf{a}(t)|} \left(\mathbf{a}(t) - proj_{\mathbf{v}(t)}\mathbf{a}(t) \right)$$
where $\mathbf{a}(t)$ and $\mathbf{v}(t)$ are the acceleration and velocity vectors associated with $\mathbf{r}(t)$. Show that the curve $\mathbf{c}(t)$ parameterized above is also given by
$$\mathbf{c}(t) = \mathbf{r}(t) + \frac{1}{\kappa(t)} \mathbf{N}(t).$$

8.2. CENTERS OF CIRCLES OF CURVATURE

6. Let $P = r(t_0)$ be a point on the curve parameterized by $\mathbf{r}(t)$. Show that the circle of curvature to the curve at P_0 can be parameterized by

$$\mathbf{r}(t_0) + \frac{1}{\kappa(t_0)}\mathbf{N}(t_0) + \mathbf{N}(t_0)\cos(t) + \mathbf{T}(t_0)\sin(t)$$

for $0 \leq t \leq 2\pi$. (Note: This formula also holds true in three-dimensions.)

7. Use the results of #6 to plot the circles of curvature to the curve given parametrically by

$$\mathbf{r}(t) = \langle \cos(t), \sin(t), t \rangle$$

for $t = 0, \pi/4$ and $\pi/2$.

8.2 Centers of Circles of Curvature

- Background material is found in Sections 10.4 of Stewart's **Multivariable Calculus: Concepts and Contexts**.

If $\mathbf{r}(t)$ parameterizes a curve in either two or three dimensions then the curve of centers of circles of curvature is given parametrically by

$$\mathbf{c}(t) = \mathbf{r}(t) + \frac{1}{k(t)}\mathbf{N}(t).$$

1. Plot the curve of centers of circles of curvature for the parabola given parametrically by

$$\mathbf{r}(t) = \langle t, t^2 \rangle.$$

2. Repeat #1 for the ellipse given parametrically by

$$\mathbf{r}(t) = \langle \cos(t), \sin(2t) \rangle.$$

3. Repeat #1 for the helix given parametrically by

$$\mathbf{r}(t) = \langle \cos(t), \sin(t), t \rangle.$$

4. Give 5 different parameterizations for the ellipse given in #2. Then find and plot $\mathbf{c}(t)$ for each of these parameterizations. What do you observe?

5. Repeat #4 for the parabola given in #1.

6. What conjecture are you prepared to make based upon your observations.

8.3 Coriolis Acceleration

- Background material is found in Section 10.4 of Stewart's **Multivariable Calculus: Concepts and Contexts**.

1. Look up the term "Coriolis force" in a good dictionary, an encyclopedia, or on the web. When did Coriolis live? What was his nationality? What was his profession?

Consider a disk of radius 1, centered at the origin, rotating about its center with constant angular speed ω. The motion of the point P_0 that is located at $(1, 0)$ at time $t = 0$ could be parametrized by

$$\mathbf{P}_0(t) = \langle \cos(\omega t), \sin(\omega t) \rangle.$$

Suppose that a bug begins moving at time $t = 0$ at constant speed c from the origin toward the point P_0 on the edge of the disk. The bug thinks it is moving in a straight line on the disk, but to a stationary observer, overhead, the bug is moving along the curved path parametrized by

$$\mathbf{r}(t) = ct\,\mathbf{P}_0(t) \quad \text{for} \quad 0 \leq t \leq 1/c.$$

2. Let $\omega = 2\pi$. Plot the path of the bug and the path of the destination point P_0 for each of $c = 1/2, 1, 2, 4,$ and 8. Add a few of the bug's velocity vectors to each plot.

3. Compute the bug's acceleration and show that it can be written as

$$\mathbf{a}(t) = 2\,\mathbf{P}'_0(t) - \omega^2 t\,\mathbf{P}_0(t).$$

Note that $\mathbf{a}(t)$ consists of a component that is parallel to the (radial) vector $\mathbf{P}_0(t)$ and a component that is orthogonal to $\mathbf{P}_0(t)$. The component that is parallel to $\mathbf{P}_0(t)$ is the **centripetal acceleration** and the component that is orthogonal to $\mathbf{P}_0(t)$ is the **Coriolis acceleration**. Redo the plots in #2 without the velocity vectors, but showing the centripetal and Coriolis acceleration vectors at several points along the bug's path.

Consider now a sphere of radius 1 centered at the origin, rotating "west-to-east" (as the Earth) about the z-axis with constant angular speed ω.

8.3. CORIOLIS ACCELERATION

Suppose that a bug travels directly "north" on the sphere with angular speed η and is located at the point $(1, 0, 0)$ on the "equator" at time $t = 0$. Then the position of the bug is given by

$$\mathbf{r}(t) = \langle \cos(\eta t) \cos(\omega t), \cos(\eta t) \sin(\omega t), \sin(\eta t) \rangle \text{ for } -\frac{\pi}{2\eta} \leq t \leq \frac{\pi}{2\eta}.$$

4. Use the information in Section 4.5 to create a wire-frame image of the unit sphere. Then add the curve

$$\mathbf{r}(t) = \left\langle \cos\left(\frac{\pi t}{2}\right) \cos(3\pi t), \cos\left(\frac{\pi t}{2}\right) \sin(3\pi t), \sin\left(\frac{\pi t}{2}\right) \right\rangle,$$

for $-1 \leq t \leq 1$, to the plot. This is the bug's path if $\omega = 3\pi$ and $\eta = 1$.

5. Compute the acceleration (in terms of ω and η) and show that it can be written as
$$\mathbf{a}(t) = \mathbf{a}_0(t) + \mathbf{a}_k(t) + \mathbf{a}_c(t)$$
where
$$\mathbf{a}_0(t) = -(\omega^2 + \eta^2) \cos(\eta t) \langle \cos(\omega t), \sin(\omega t), 0 \rangle$$
$$\mathbf{a}_k(t) = \langle 0, 0, -\eta^2 \sin(\eta t) \rangle$$
and
$$\mathbf{a}_c(t) = 2\omega\eta \sin(\eta t) \langle \sin(\omega t), -\cos(\omega t), 0 \rangle.$$
The vector $\mathbf{a}_c(t)$ is the **Coriolis acceleration on the unit sphere**.

6. Show that:

    (a) $\mathbf{a}_0(t)$ is parallel to the xy-plane and points toward the z-axis for $\frac{-\pi}{2\eta} < t < \frac{\pi}{2\eta}$.

    (b) $\mathbf{a}_k(t)$ is parallel to the z-axis when $t \neq 0$, pointing up when $\frac{-\pi}{2\eta} < t < 0$ and down when $0 < t < \frac{\pi}{2\eta}$.

    (c) $\mathbf{a}_c(t)$ is orthogonal to $\mathbf{a}_0(t)$, $\mathbf{a}_k(t)$, and the position vector $\mathbf{r}(t)$.

7. The plane curve parametrized by

$$\mathbf{q}(t) = \left\langle \cos\left(\frac{\pi t}{2}\right) \cos(3\pi t), \cos\left(\frac{\pi t}{2}\right) \sin(3\pi t) \right\rangle, \quad -1 \leq t \leq 1$$

is the bug's path plotted in problem 4, as viewed from a point on the z-axis high above the sphere.

(a) Create a parametric plot of this curve, using a $[3.5, 3.5] \times [-1.5, 1.5]$ plot window. Take note of which part of the curve corresponds to the bug's path in the "southern hemisphere", and which part corresponds to the bug's path in the "northern hemisphere".

(b) The Coriolis acceleration vectors, as viewed from a point on the z-axis high above the sphere, are the two-dimensional vectors

$$\mathbf{a}_c(t) = 3\pi^2 \sin\left(\frac{\pi t}{2}\right) \langle \sin(3\pi t), -\cos(3\pi t) \rangle$$

in the xy-plane. Plot six of these vectors on the "southern hemisphere" portion of the curve plotted in (a). Do these vectors point to the "east" (right) or to the "west" (left) of the bug's path? (Remember that the bug is moving upward, toward your eye.)

(c) Now plot six Coriolis acceleration vectors on the "northern hemisphere" portion of the curve plotted in (a). Do these vectors point to the "east" (right) or to the "west" (left) of the bug's path?

8. Show that the Coriolis acceleration points in the same direction (east or west) regardless of whether the bug travels south-to-north or north-to-south in each of the two hemispheres, respectively.

9. Suppose that a river flows south-to-north or north-to-south in the southern hemisphere. What effect might the Coriolis acceleration of the water have? What about in the northern hemisphere?

8.4 An Ant on a Helix

- Background material is found in Sections 10.1 - 10.3 of Stewart's **Multivariable Calculus: Concepts and Contexts**.

Suppose that an ant walks along a helical path on the surface of a transparent cylinder $x^2 + y^2 = 1$ such that its position vector at time t is

$$\mathbf{r}(t) = \langle \cos(t), \sin(t), t \rangle.$$

A light source is located on the z-axis at the point $(0, 0, 8\pi)$, causing the ant to cast a shadow on the xy-plane while $0 \leq t \leq 8\pi$.

1. Let $rxy(t)$ denote the path of the ant's shadow in the xy-plane. Without finding a parameterization for $rxy(t)$, explain why
$$\lim_{t \to 8\pi^-} |rxy(t)| = \infty.$$

2. Find the parameterization for $rxy(t)$ and plot its graph. Create a second, three-dimensional plot showing the helix and the path of the shadow on the xy-plane..

3. Find the length of the curve traced by the ant's shadow while $0 \leq t \leq 4\pi$.

8.5 3D Graphics

- Background material is found in Sections 9.1–9.5 and 10.1 of Stewart's **Multivariable Calculus: Concepts and Contexts**.

A fundamental problem in computer graphics is the projection of a point in three-dimensional space onto a *view-plane* and then determining two-dimensional *view-plane coordinates* for the projected point. We will assume that our "eye" is located at a point E and looks directly toward the origin. The view-plane is orthogonal to the line of sight between E and the origin, and the point C is the intersection of the view-plane with that line of sight.

1. Give a sketch associated with the description above.

2. Let P be a point in space such that the half line through E with direction vector $\overrightarrow{EP}$ intersects the view plane. Justify each of the following steps in the derivation of a formula for Q, the **view-plane projection** of P.

    (a) Let $X = \langle x, y, z \rangle$. An equation of the view-plane is $X \cdot E = \mu E \cdot E$, where $0 < \mu < 1$ and $\mu E = C$.

    (b) A parametrization of the line of sight between E and P is given by
    $$r(t) = P + t(E - P).$$

(c) The point where the line of sight between E and P intersects the view-plane is given by
$$Q = P + \frac{P \cdot E - \mu E \cdot E}{P \cdot E - E \cdot E}(E - P).$$

3. Let $\mathbf{k} = \langle 0, 0, 1 \rangle$ and define
$$\mathbf{v} = \frac{1}{|E - proj_E \mathbf{k}|}(E - proj_E \mathbf{k}).$$
Then set
$$\mathbf{u} = \frac{1}{|E \times \mathbf{v}|} E \times \mathbf{v}.$$
Explain why the vectors $\mathbf{u}$ and $\mathbf{v}$ form a natural set of orthogonal coordinate axes on the view plane if they are placed with their initial ends at the point C.

4. We can now use $\mathbf{u}$ and $\mathbf{v}$ to define view plane coordinates for points projected onto the view plane. Let P be a point in space such that the half line through E with direction vector $\overrightarrow{EP}$ intersects the view plane. Let Q be the view plane projection of P. We define the **view plane coordinates** of the projection of P onto the view plane as
$$(\mathbf{Q} \cdot \mathbf{u}, \mathbf{Q} \cdot \mathbf{v}).$$
Explain why this definition is made by showing that
$$Q = C + (\mathbf{Q} \cdot \mathbf{u})\, u + (\mathbf{Q} \cdot \mathbf{v})\, v.$$

5. Let the eye point E be given by $E = (10, 10, 20)$ and suppose the view plane is given by
$$x + y + 2z = 5.$$
Sketch the view plane projection of the curve parameterized by
$$r(t) = \langle \cos(t), \sin(2t), \cos(3t) \rangle$$
by plotting the view plane coordinates of the projection of each point on the curve (as a two dimensional plot).

8.6 Least Squares and Curve Fitting

- Background material is found in Section 11.7 of Stewart's **Multivariable Calculus: Concepts and Contexts**. In particular, see Stewart's exercise 45.

Fitting a Line. Suppose that twenty-four fish of a certain species are measured and weighed, resulting in the following list of ordered pairs containing the length and weight measurements (in inches and pounds, respectively) for each fish.
$(9.6, 6.4)$, $(16.4, 11.7)$, $(10.4, 6.6)$, $(11.1, 7.9)$, $(11.8, 10.)$, $(11.7, 8.7)$, $(9.5, 8.3)$, $(19.2, 16.2)$, $(11.3, 10.3)$, $(17.7, 15.3)$, $(11.5, 10.)$, $(10.3, 8.4)$, $(18.2, 12.9)$, $(8., 7.7)$, $(5.5, 3.4)$, $(14.4, 9.5)$, $(14.9, 12.4)$, $(20.9, 17.7)$, $(20.7, 16.)$, $(16.5, 14.3)$, $(5.5, 5.5)$, $(19.7, 14.8)$, $(14.1, 9.5)$, $(12.1, 8.4)$
The following is a *scatterplot* of the data in a $[-1, 25] \times [-1, 17]$ window.

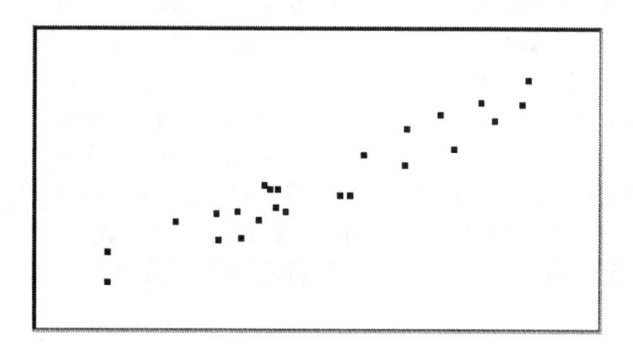

Observing that there seems to be a linear trend in the scatterplot, we would like to determine the equation of the straight line that *"best fits"* the data. So we wish to find values of a and b so that the graph of $g(x) = ax + b$ is as close as possible to our data points. We will do this by minimizing the **least-squares error**, defined here by

$$\text{LSE}(a, b) = \sum_{i=1}^{24} (g(x_i) - y_i)^2,$$

where (x_i, y_i) denotes the ith data point.

1. Use **DERIVE** to create the point plot above.

2. Define the model function
$$g(a, b, x) = ax + b$$
and least-squares error function, and then compute the partial derivatives of **LSE** with respect to **a** and **b**.

3. Solve for the critical point and plot the resulting line along with the scatterplot of the data.

4. Predict the weight of such a fish if it were 24 inches long.

5. Verify that the average of the values $g(x_1), ..., g(x_{24})$ is the same as the average of the values $y_1, ..., y_{24}$.

A Quadratic Fit. A biologist is interested in the relationship between air temperature and the presence of *sandgnats* in Savannah, Georgia. Sandgnats are particularly pernicious pests that can practically preclude outdoor activity by humans during certain times of the year. At a certain location prone to sandgnat swarms, the biologist makes temperature measurements and sandgnat counts at sunset on twenty-five different days during the months of April and May. The measurements were as follows.
(80, 141), (66, 196), (65, 193), (67, 203), (78, 162), (78, 155), (52, 92), (66, 199), (58, 165), (83, 108), (49, 44), (84, 100), (77, 174), (56, 131), (86, 54), (70, 188), (73, 180), (79, 156), (77, 183), (64, 195), (76, 193), (57, 171), (86, 79), (67, 225), (83, 113)

6. Enter the given data into **DERIVE** as a matrix named **gnats**. Then create a scatterplot of the data in a $[40, 95] \times [0, 230]$ window.

You should notice that the shape of the scatterplot suggests a roughly quadratic relationship between temperature and the number of sandgnats. So we wish to find values of a and b so that the graph of
$$g(x) = ax^2 + bx + c$$
is as close as possible to our data points. We will do this by minimizing the **least-squares error**, defined here by
$$\mathbf{LSE}(a, b, c) = \sum_{i=1}^{25} (g(x_i) - y_i)^2,$$

where (x_i, y_i) denotes the ith data point. (Even though the model function is nonlinear, this is still a *linear* least-squares problem, since the gradient of the least-squares error will be a linear function of a, b, and c.)

7. By analogy with problems 2 and 3, define a model function **g(a,b,c,x)** and least-squares error **LSE(a, b, c)**. Then solve for the critical point of the least squares error.

8. Plot the resulting parabola along with the scatterplot of the data.

9. Use the model to estimate the temperature that sandgnats most enjoy.

8.7 Classifying Critical Points

- Background material is found in Section 11.7 of Stewart's **Multivariable Calculus: Concepts and Contexts**.

Quadratic Functions. Any quadratic function f of n variables can be expressed in "vector form" as

$$f(u) = \frac{1}{2} u H u^T + a \cdot u + b,$$

where $u = \langle u_1, \ldots, u_n \rangle$ is the row vector containing the variables, a is a vector of length n, b is a number, and H is the $n \times n$ **Hessian** matrix of f, whose ij^{th} entry is

$$[H]_{i,j} = \frac{\partial^2 f}{\partial u_i \partial u_j}.$$

For example, with

$$H = \begin{bmatrix} 2 & 3 \\ 3 & 5 \end{bmatrix}, \quad u = \langle x, y \rangle, \quad a = \langle 7, 5 \rangle, \text{ and } b = 5,$$

we have

$$u H u^T = \langle x, y \rangle \begin{bmatrix} 2 & 3 \\ 3 & 5 \end{bmatrix} \begin{bmatrix} x \\ y \end{bmatrix} = \langle x, y \rangle \begin{bmatrix} 2x + 3y \\ 3x + 5y \end{bmatrix} = 2x^2 + 6xy + 5y^2,$$

and so

$$\frac{1}{2} u H u^T + a \cdot u + b = \frac{1}{2} \left(2x^2 + 6xy + 5y^2 \right) + 7x + 5y + 5.$$

On the other hand, given a quadratic function f, such as

$$f(x,y,z) = x^2 + 2y^2 + 3z^2 - yz + xy + x - y + 3,$$

the Hessian matrix of f can be found as follows:

$$H = \begin{bmatrix} f_{xx} & f_{xy} & f_{xz} \\ f_{yx} & f_{yy} & f_{yz} \\ f_{zx} & f_{zy} & f_{zz} \end{bmatrix} = \begin{bmatrix} 2 & 1 & 0 \\ 1 & 4 & -1 \\ 0 & -1 & 6 \end{bmatrix}.$$

Notice that each row of the Hessian matrix is essentially the gradient of an entry in the gradient of f. So in a sense, the Hessian matrix of f is the "gradient of the gradient" of f. Indeed, sometimes the notation $\nabla^2 f$ is used to denote the Hessian matrix of f. Unfortunately, this notation also coincides with the notation used by physicists for the Laplacian. What ever notation is used, the Hessian is the second derivative of f. For this reason, we will use the alternate notation f'' for the Hessian.

Positive Definiteness. A matrix H is said to be **positive definite** if

$$uHu^T > 0 \text{ for all nonzero vectors } u.$$

If the matrix H is positive definite, then H is invertible, and the function

$$f(u) = \frac{1}{2} uHu^T + a \cdot u + b$$

has a single critical point at which f attains a global minimum. The following theorem provides a means for determining whether a matrix is positive definite. Its proof can be found in many linear algebra texts.

Theorem 66 *The matrix H is positive definite if and only if the determinants of all of the principal minors of H are positive.*

Any $n \times n$ matrix H has n principal minors $M_1, M_2, \ldots, M_n$, where M_k is the $k \times k$ matrix obtained by deleting all but the first k rows and k columns of H. (Thus $M_n = H$.)

For instance, suppose that H is the 4×4 matrix given by

$$H = \begin{bmatrix} 1 & -1 & 0 & 0 \\ -1 & 2 & -1 & 0 \\ 0 & -1 & 3 & -1 \\ 0 & 0 & -1 & 4 \end{bmatrix}.$$

8.7. CLASSIFYING CRITICAL POINTS

The first three principal minors of H are given by

$$M_1 = [1], \quad M_2 = \begin{bmatrix} 1 & -1 \\ -1 & 2 \end{bmatrix}, \quad M_3 = \begin{bmatrix} 1 & -1 & 0 \\ -1 & 2 & -1 \\ 0 & -1 & 3 \end{bmatrix}$$

Of course, as stated in the theorem above. our concern is with the determinants of the principal minors.

1. For the quadratic function

$$f(x, y, z) = 3x^2 - 3xy + 2y^2 - 2yz + z^2 + x - y + z,$$

   compute the Hessian matrix H and show that H is positive definite. Then find the critical point and the minimum value of f.

2. For each of the following functions, all of which have a critical point at the origin, compute the Hessian matrix H and determine whether H is positive definite.

   (a) $f(x, y) = x^2 + 5y^2 - 7xy$
   (b) $f(x, y, z) = x^2 + 5y^2 + z^2 - yz - 4xy$
   (c) $f(x, y, z) = x^2 + 4y^2 + z^2 - yz - 4xy$
   (d) $f(w, x, y, z) = w^2 + x^2 + 5y^2 - yz - wz$

3. Let $f(x, y, z) = x^2 + 3y^2 + 5z^2 - 3yz - cxy$. Describe all values of c for which f has a single critical point and a global minimum at $(0, 0, 0)$.

4. A matrix H is said to be **negative definite** if

$$uHu^T < 0 \text{ for all nonzero vectors } u.$$

   If the matrix H is negative definite, then H is invertible, and the function

$$f(u) = \frac{1}{2} uHu^T + a \cdot u + b$$

   has a single critical point at which it attains a global *maximum*.

   (a) Argue that H is negative definite if and only if $-H$ is positive definite.

(b) Verify that the Hessian matrix of
$$f(x,y,z) = -x^2 - 2y^2 - 3z^2 + 2xy - yz$$
is negative definite. Then find the critical point and the maximum value of f.

5. If the Hessian matrix H of f is invertible and *indefinite*, i.e., neither positive definite nor negative definite, then f has neither a maximum nor a minimum value at its critical point. In this case, the critical point is called a **saddle point**.

(a) Describe all values of c such that $(0,0)$ is a saddle point of
$$f(x,y) = x^2 + 2y^2 + cxy.$$
Then create a surface plot and a contour plot of the surface $z = f(x,y)$ for one such c.

(b) Describe all values of c such that $(0,0)$ is a saddle point of
$$f(x,y) = x^2 - y^2 + cxy.$$
Then create a surface plot and a contour plot of the surface $z = f(x,y)$ for one such c.

General Nonlinear Functions. If f is a twice-differentiable function of n variables—i.e., a twice-differentiable function of a vector $x = \langle x_1, x_2, ..., x_n \rangle$ of length n—then the quadratic approximation of f at a point P is given by
$$Q_P(x) = f(P) + \nabla f(P) \cdot (x-P) + \frac{1}{2}(x-P)f''(P)(x-P)^T$$
where $f''(P)$ is the Hessian matrix of f at P, whose ij^{th} entry is
$$[f''(P)]_{i,j} = \frac{\partial^2 f}{\partial x_i \partial x_j}\big|_{x=P}.$$

The behavior of f at a critical point P is essentially determined by the behavior of its quadratic approximation at P. More precisely, the following theorem is true:

8.8. OPTIMIZATION

Theorem 67 *Suppose that P is a critical point of f, i.e., $\nabla f(P) = \vec{0}$.*

a. *If $f''(P)$ is positive definite, then f has a local minimum at P.*

b. *If $f''(P)$ is negative definite, then f has a local maximum at P.*

c. *If $f''(P)$ is invertible and indefinite, then P is a saddle point.*

6. Given the function
$$f(x, y, z) = x \sin x + z^2 \cos(yz) + yz + 2x^2 + xy + y^2 e^{xz},$$
verify that $(0, 0, 0)$ is a critical point. Then show that f has a local minimum at $(0, 0, 0)$.

7. Given the function
$$f(x, y, z) = x^4 + y^3 + z^2 + xy + yz + xz,$$

(a) Locate all the critical points of f.

(b) For each of the (real) critical points, compute the Hessian matrix and determine whether it is positive definite, negative definite, or indefinite. Use this information to classify each critical point.

8.8 Optimization

Background material is found in Sections 11.6 and 11.7 of Stewart's **Multivariable Calculus: Concepts and Contexts**.

Given a function $f(x, y)$, its **quadratic approximation** at (a, b) is

$$\begin{aligned} Q_{(a,b)}(x, y) &= f(a, b) + f_x(a, b)(x - a) + f_y(a, b)(y - b) \\ &+ \frac{1}{2}(f_{xx}(a, b)(x - a)^2 + 2f_{xy}(a, b)(x - a)(y - b) \\ &+ f_{yy}(a, b)(y - b)^2). \end{aligned}$$

Think of this as the function whose graph is the paraboloid that best approximates the surface $z = f(x, y)$ at $(a, b, f(a, b))$. The preceding formula can be expressed more concisely as

$$\begin{aligned} Q_{(a,b)}(x, y) &= f(a, b) + \nabla f(a, b) \cdot \langle x - a, y - b \rangle \\ &+ \frac{1}{2} \langle x - a, y - b \rangle f''(a, b) \begin{bmatrix} x - a \\ y - b \end{bmatrix} \end{aligned}$$

where $f''(a, b)$ is the Hessian matrix of f at (a, b), defined by

$$f''(a,b) = \begin{bmatrix} f_{xx}(a,b) & f_{xy}(a,b) \\ f_{yx}(a,b) & f_{yy}(a,b) \end{bmatrix}.$$

Note that the products in the third term of $Q_{(a,b)}(x, y)$ are matrix products. More generally, if f is a function of n variables—i.e., a function of a (row) vector x of length n—then the quadratic approximation of f at the point P is given by

$$Q_P(x) = f(P) + \nabla f(P) \cdot (x - P) + \frac{1}{2}(x - P) f''(P)(x - P)^T.$$

1. Find the quadratic approximation $Q_{(1,1)}(x, y)$ of

$$f(x, y) = \sin(x - y)^2 + \sin^2(x + y - 2)$$

at the point $(1, 1)$, where it happens to have a locally minimum value. (Why?) Plot each of the following pairs of surface traces and comment on what you observe.

   (a) $f(x, 1)$ and $Q_{(1,1)}(x, 1)$ for $-1 \leq x \leq 3$
   (b) $f(1, y)$ and $Q_{(1,1)}(1, y)$ for $-1 \leq y \leq 3$
   (c) $f(x, x)$ and $Q_{(1,1)}(x, x)$ for $-1 \leq x \leq 3$

2. Repeat #1 for the function

$$f(x, y) = x^2 + 2\sin\left((x - y)^2\right) + 3\sin^2(x + y - 2),$$

which attains a local minimum value *near*, but not at, $(1, 1)$.

Newton's Method for Minimization. Suppose that f is a quadratic function (of any number of variables) that attains a minimum value at a unique point P. In general, the problem of finding this point is simple, though not necessarily easy. This is because the gradient will be a linear function, and so finding the critical point amounts to solving a *linear* system of equations. An iterative method for locating local minimizers of smooth, non-quadratic functions—based on quadratic approximation—takes advantage of this. The method is **Newton's Method for Minimization** and is described as follows:

For $k = 0, 1, 2, \ldots$

8.8. OPTIMIZATION

   i. Find the minimizer of the quadratic approximation of f at x_k.

   ii. Let this new point be x_{k+1}.

It turns out that the minimizer of the quadratic approximation of f at x_k—assuming there is one—is easily found. It is
$$x_k + x_k^T, \text{ where } f''(x_k)s_k = -\nabla f(x_k)^T.$$
In fact, the method can be restated as the following algorithm:
For $k = 0, 1, 2, \ldots$

   i. Form the matrix $M_k = \nabla^2 f(x_k)$ and the row vector $q_k = \nabla f(x_k)$.

   ii. Solve $M_k s_k = -q^T$ for the column vector s_k.

   iii. Let $x_{k+1} = x_k + s_k^T$.

This is called Newton's Method for Minimization, because—for minimization problems in which the Hessian matrix is always positive definite—it is equivalent to Newton's Method for finding points where the gradient of f is $\vec{0}$. (See the next project on Newton's Method for systems of nonlinear equations.)

Consider the quadratic approximation $Q_{(a,b)}(x, y)$ of a function $f(x, y)$ of two variables given earlier.

3. Show that the gradient of $Q_{(a,b)}$ is
$$\nabla Q_{(a,b)}(x, y) = \begin{pmatrix} f_x(a,b) + f_{xx}(a,b)(x-a) + f_{xy}(a,b)(y-b) \\ f_y(a,b) + f_{xy}(a,b)(x-a) + f_{yy}(a,b)(y-b) \end{pmatrix}^T.$$

4. Show that $\nabla Q_{(a,b)}(x,y) = \vec{0}$ when (x,y) satisfies the equation
$$\begin{bmatrix} f_{xx}(a,b) & f_{xy}(a,b) \\ f_{xy}(a,b) & f_{yy}(a,b) \end{bmatrix} \begin{bmatrix} x-a \\ y-b \end{bmatrix} = -\begin{bmatrix} f_x(a,b) \\ f_y(a,b) \end{bmatrix},$$
which may also be written as
$$f''(a,b)\langle x-a, y-b \rangle^T = -\nabla f(a,b)^T.$$
Conclude that if (x, y) satisfies this equation and $f''(a, b)$ is *positive definite* (see the *Classifying Critical Points* project), then $Q_{(a,b)}(x, y)$ attains an absolute minimum value at (x, y). Describe how this result gives rise to the algorithm given above in the two-variable case.

5. The function
$$f(x,y) = x^2 + 2\sin\left((x-y)^2\right) + 3\sin^2(x+y-2)$$
has a local minimizer near $(1,1)$. Use Newton's method to locate the minimizer to six decimal-place accuracy. Also compute the values of f and ∇f at the result.

6. The function
$$f(x,y) = e^{-\sqrt{x^2+3y^2}} \cos\sqrt{x^2+y+2y^2}$$
has a local minimizer in the rectangle $1 \leq x \leq 3$, $-1 \leq y \leq 2$. Use a surface plot (or a contour plot) to obtain a rough estimate of the minimizer. Then find the minimizer to six decimal-place accuracy. Compute the values of f and ∇f at the result.

7. The function
$$f(x,y,z) = (y-x+z)^2 + (3-x-y^2)^2 + (3-x-y^2)^2 + (6-x-y^2-z^3)^2$$
has a local minimizer near $(2, 1, 1.5)$. Compute six Newton steps to find an approximation to that minimizer. Compute the values of f and ∇f at the resulting approximation.

8.9 The Gradient Descent Method

- Background material is found in Section 11.7 of Stewart's **Multivariable Calculus: Concepts and Contexts**.

If f is a real valued, differentiable function of n variables and $P \in R^n$ then $\nabla f(P)$ points in the direction of greatest increase of the function f at the point P, and $-\nabla f(P)$ points in the direction of greatest decrease at the point P. This leads to an interesting method which can be used to search for places where f has a local or global minimum. This method is referred to at the **Gradient Descent Method**, and the algorithm is shown below:

(i) Give a point P as a guess of a point where f has a local (or global) minimum. Assume $\nabla f(P) \neq \vec{0}$.

8.9. THE GRADIENT DESCENT METHOD

(ii) Form the vector
$$v = \frac{1}{\|\nabla f(P)\|} \nabla f(P),$$
the half line
$$P - tv, \text{ for } t \geq 0,$$
and define a new function
$$g(t) = f(P - tv), \text{ for } t \geq 0$$
(the restriction of f to the half line).

(iii) Find the smallest value $t_0 > 0$ such that g has a local minimum at t_0 and form the point
$$Q = P - t_0 v.$$

If a **stopping criterion** is met (see below) then exit the algorithm. Otherwise, replace P with Q in step (i) and repeat the process.

The stopping criterion referred to above could be either t_0 is sufficiently small (and consequently Q is essentially P), or $\nabla f(Q)$ is small (in norm).

For example, consider the problem of finding a point where
$$f(x, y) = x^2 + xy + 3y^2 + 2x + \exp(x - 2y)$$
has a local minimum. Let's apply the gradient descent method with an initial guess of $(-1, -1)$. You should be able to compute
$$\nabla f(x, y) = \left(2x + y + 2 + e^{x-2y}, x + 6y - 2e^{x-2y}\right)$$
$$\nabla f(-1, -1) = \langle 1.7183, -12.437\rangle$$
and
$$g(t) = f(-1 - t(1.7183), -1 - t(-12.437)).$$
That is
$$g(t) = 3 - 85.341t + 445.62t^2 + \exp(1 - 26.592t).$$
A graph of $g(t)$ is shown below.

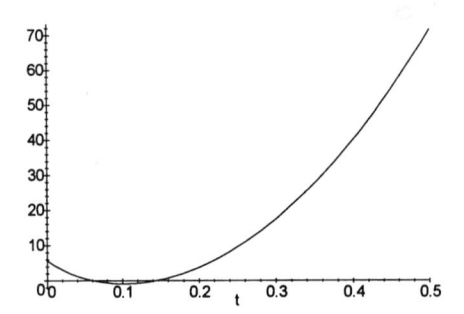

We can see that g has a local minimum near $t = 0.1$. Furthermore, if we numerically solve
$$g'(t) = 0$$
then we find $t_0 = .10125$. Consequently, the new approximation of for a local minimum of f given by the gradient descent method is

$$(-1 - .10125\,(1.7183), -1 - .10125\,(-12.437)) = (-1.174, .25925).$$

1. Use **DERIVE** to continue the process above until the approximate for the point at which f has a local minimum is accurate to 4 decimal places.

2. Use the final point obtained in #1 as an initial guess for the **NEWTONS** command to solve $\vec{\nabla} f(x,y) = \vec{0}$. Then use the second derivative test to verify that this point is a local minimum.

3. Repeat the process above for the function
$$f(x,y) = x^4 + y^4 + x^2 y^2 + x^2 + x - 2y.$$

4. Give a graphical interpretation of the gradient descent process for the function
$$f(x,y) = x^2 + 3y^2 + x - y$$
starting from a guess of $(1,1)$ by sketching a contour plot of f and then showing the path taken by gradient descent in its search for the minimum value of f.

8.10 Newton's Method

- Background material is found in Sections 11.6 and 11.7 of Stewart's **Multivariable Calculus: Concepts and Contexts**.

Recall Newton's Method for finding roots of a function f of one variable:

$$x_{k+1} = x_k - \frac{f(x_k)}{f'(x_k)}, \quad k = 0, 1, 2, \ldots,$$

which is derived by solving the linearized problem at x_k, given by

$$f(x_k) + f'(x_k)(x - x_k) = 0.$$

The analogous method for a system of equations $F(\mathbf{x}) = \vec{0}$, where F is a function from R^n to R^n is derived in the same way. The linearized problem is

$$F(\mathbf{x}_k) + J_F(\mathbf{x}_k)(\mathbf{x} - \mathbf{x}_k) = \vec{0},$$

where $J_F(\mathbf{x}_k)$ is Jacobian matrix of F at $\mathbf{x}_k$ (i.e. the derivative of F at $\mathbf{x}_k$). Here we are thinking of $\mathbf{x}$, $\mathbf{x}_k$, and $F(\mathbf{x}_k)$ as column vectors. The ij^{th} entry of $J_F(\mathbf{x}_k)$ is the partial derivative of the ith component of F with respect to the jth variable in $\mathbf{x}$. For example, the system of equations

$$\left\{ \begin{array}{r} x + 2y - z = 0 \\ x^2 y + z = 0 \\ x + yz + 2 = 0 \end{array} \right\}$$

would be viewed as

$$F(\mathbf{x}) = \vec{0},$$

where

$$\mathbf{x} = \begin{bmatrix} x \\ y \\ z \end{bmatrix} \text{ and } F(\mathbf{x}) = \begin{bmatrix} x + 2y - z \\ x^2 y + z \\ x + yz + 2 \end{bmatrix}.$$

The Jacobian matrix here is easily seen to be

$$J_F(\mathbf{x}) = \begin{bmatrix} 1 & 2 & -1 \\ 2xy & x^2 & 1 \\ 1 & z & y \end{bmatrix}$$

1. Show that solving the linearized problem
$$F(\mathbf{x}_k) + J_F(\mathbf{x}_k)(\mathbf{x} - \mathbf{x}_k) = \vec{0}$$
for **x** produces
$$\mathbf{x}_{k+1} = \mathbf{x}_k - J_F(\mathbf{x}_k)^{-1} F(\mathbf{x}_k), \quad k = 0, 1, 2, \ldots$$
provided $J_F(\mathbf{x}_k)^{-1}$ exists. This is Newton's method for systems of equations.

Because it is more efficient than computing the inverse of the Jacobian matrix at each step, we will restate Newton's Method in two "sub-steps" as:
For $k = 0, 1, 2, \ldots,$

i. Solve $J_F(\mathbf{x}_k)\mathbf{s}_k = -F(\mathbf{x}_k)$ for $\mathbf{s}_k$;

ii. Set $\mathbf{x}_{k+1} = \mathbf{x}_k + \mathbf{s}_k$.

Note: DERIVE has a built in command **NEWTONS** which performs Newton's method.

2. Use Newton's method to find a solution of the system
$$x^2 + y^2 + z^2 = 21/64$$
$$8xz - \sin^2 \pi y = 0$$
$$2x + 4y + 8z = 3.$$

Begin the iteration at $(1, 0, 0)$ and compute six iterates. Do the approximations reveal the exact solution? What is it?

3. Let
$$f(x, y, z, w) = x + z - x^2 y - y^2 z + zw + w^2.$$

(a) Locate a critical point of f by applying Newton's Method to ∇f.

(b) f actually has three (real) critical points. Find them all with Newton's Method.

4. Consider the function from R^2 to R^2 given by
$$F(x, y) = (x^3 - y^2, x^2 - y^3).$$

Use Newton's Method to find a nonzero *fixed point* of F; that is, a point $(x, y) \neq (0, 0)$ such that $F(x, y) = (x, y)$.

8.11 Balancing a Region

- Related material is found in Section 12.7 of Stewart's **Multivariable Calculus: Concepts and Contexts**.

Suppose E is a bounded region in three-dimensional space, and suppose we can manipulate E. For example, if E is defined as

$$E = \{(x, y, z) \mid x^2 + y^2 + z^2 \leq 1\}$$

(the unit disk) then we can think of E as a ball, and we might imagine rolling or throwing this ball. Notice that in this case, if we place E on a table (perfectly horizontal), then it will balance on the point of contact (regardless of which point of the ball touches the table). This is certainly not true for other choices of E. As an example, consider a slight modification given by

$$E = \left\{(x, y, z) \mid x^2 + \frac{y^2}{4} + \frac{z^2}{3} \leq 1\right\}.$$

This time E is an ellipsoidal ball.

1. If E is placed on a table then depending upon the point of contact we E might not balance. Find all points (in xyz-coordinates) on the boundary of E on which E can balance.

In general, this problem can be found in two steps. First we find the center of mass of E. Then we find the points P on the boundary of E where the normal line to the boundary through P passes through the center of mass.

2. Consider the region given by

$$E = \{(x, y, z) \mid x^2 + y^2 - 1 \leq z \leq 8 - 2x^2 - 2y^2\}.$$

Find all of the points of contact on the boundary of E on which E can balance.

3. Repeat #2 for the region given by

$$E = \{(x, y, z) \mid x^2 + y^2 - 1 \leq z \leq 8 - 2x^2 - y^2\}.$$

4. Consider the balancing points found in problems 1, 2 and 3. Which of these points so you think will be stable to small perturbations? By the term "stable", we mean that if E is placed with a point of contact near one of the balancing points, then E might "rock", but will stay close to the balancing point. Give reasons for your conclusions.

8.12 Velocity Fields and Steady Flow

- Background material is found in Section 13.1 of Stewart's **Multivariable Calculus: Concepts and Contexts**.

A continuous vector field determines a family of **flow lines**. These are smooth curves that are everywhere tangent to the vector field.

In the context of fluid flow, the vector field of interest is the velocity field of the fluid at each point in its domain. When the velocity field is independent of time, the flow is said to be a steady flow, and flow lines are usually called **streamlines**. So imagine a steady flow of fluid particles moving in two dimensions in such a way that velocity (independent of time) is a function of location (x, y) given by

$$\mathbf{V}(x, y) = \langle u(x, y), v(x, y) \rangle .$$

Then the streamlines are curves of the form $(x(t), y(t))$ which satisfy the system of differential equations

$$\begin{aligned} x'(t) &= u(x(t), y(t)) \\ y'(t) &= v(x(t), y(t)) \end{aligned}$$

If $(x(0), y(0)) = (x_0, y_0)$ then (x_0, y_0) is referred to as an initial condition for the streamline $(x(t), y(t))$.

Before starting the exercises below, use the help facility in **DERIVE** to learn how to plot solution curves to systems of differential equations.

1. Plot the vector field

$$\mathbf{V}(x, y) = \left\langle \frac{y}{x^2 + y^2}, \frac{x}{x^2 + y^2} \right\rangle$$

8.13. INCOMPRESSIBLE POTENTIAL FLOW

on the rectangle $-4 \leq x \leq 4$, $-2 \leq y \leq 2$ along with stream lines associated with the initial conditions

$$(x_0, y_0) = (1, 0), (2, 0)$$

for $0 \leq t \leq 2\pi$. What kind of curves do the stream lines appear to be? Prove your conjecture.

2. Plot the vector field

$$\mathbf{V}(x, y) = \left\langle 2 + \frac{x}{x^2 + y^2}, \frac{x}{x^2 + y^2} \right\rangle$$

on the square described by $-2 \leq x \leq 2$, $-2 \leq y \leq 2$ along with stream lines associated with the initial conditions

$$(x_0, y_0) = (-2, -1), (-2, -.5), (-2, -.1) (-2, .5)$$
$$(-2, 1), (.01, -.1), (.01, .1)$$

for $0 \leq t \leq 2$.

8.13 Incompressible Potential Flow

- Some background material is found in Sections 13.1 and 13.5 of Stewart's **Multivariable Calculus: Concepts and Contexts**.

Given a steady two-dimensional fluid flow with velocity given by

$$\mathbf{V}(x, y) = \langle u(x, y), v(x, y) \rangle,$$

its **rotation** is defined as

$$\omega = \text{rot}(V) = v_x - u_y.$$

For a steady three-dimensional fluid flow with velocity field V, the **vorticity** of the flow is defined to be the curl of the velocity

$$\omega = \nabla \times \mathbf{V}.$$

In either case, if $\omega = 0$ (or $\vec{\omega} = \vec{0}$) at all points within the flow, then the flow is said to be **irrotational**.

To imagine an irrotational flow, imagine one in which a very short straw will move without rotating; i.e., without changing its angle of orientation with respect to any coordinate axis. It turns out that irrotationality is a necessary and sufficient condition for a flow to be a **potential flow**, that is, a flow whose velocity field $\mathbf{V}$ is the gradient of some potential function Φ.

1. Check that the 2D flow whose velocity field is
$$\mathbf{V}(x, y) = \langle 2xy - 2y^2, x^2 - 4xy \rangle$$
is irrotational and therefore a potential flow. Find the potential function Φ. Plot the flow field on the square $-2 \leq x \leq 2$, $-2 \leq y \leq 2$.

2. Check that the 3D flow whose velocity field is
$$\mathbf{V}(x, y) = \langle 3x^2 y + 3z^2, x^3 - z, -y + 6xz \rangle$$
is irrotational and therefore a potential flow. Find the potential function Φ.

The flow of an **incompressible** fluid with velocity field $\mathbf{V}$ must satisfy the continuity equation
$$\text{div}(\mathbf{V}) = 0.$$
(The flow of an incompressible fluid is said to be **divergence free**. Water is an example of an essentially incompressible fluid. Air is an example of a fluid that is compressible.) Thus, a steady incompressible and irrotational flow is one whose velocity field $\mathbf{V}$ satisfies:
$$\text{div}(\mathbf{V}) = 0 \text{ and } \text{rot}(\mathbf{V}) = 0 \text{ for 2D flow;}$$
$$\text{div}(\mathbf{V}) = 0 \text{ and } \text{curl}(\mathbf{V}) = \vec{0} \text{ for a 3D flow.}$$
Irrotationality implies that $\mathbf{V}$ has a potential function Φ. This together with the continuity equation, implies that Φ satisfies
$$\nabla \cdot \mathbf{V} = \nabla \cdot \nabla \Phi = 0;$$
that is, the velocity potential Φ satisfies Laplace's equation:
$$\nabla^2 \Phi = 0.$$
(Note that some textbooks use the notation $\Delta \Phi$ instead of $\nabla^2 \Phi$.) Moreover, the gradient of any function Φ that satisfies Laplace's equation satisfies the conditions to be the velocity field of an incompressible irrotational fluid flow.

8.14. REFLECTING POINTS

3. Show that
$$\Phi(x, y) = 2x + \frac{1}{2}\log(x^2 + y^2)$$
satisfies Laplace's equation for all $(x, y) \neq (0, 0)$. Give the associated velocity field. Plot this velocity field. (This potential function models a uniform left-to-right flow superimposed over a "source" located at the origin. Imagine a river flowing over a spring.)

4. Consider the vector field
$$\mathbf{V}(x, y) = \left\langle 1 - \frac{x^2 - y^2}{(x^2 + y^2)^2}, -\frac{2xy}{(x^2 + y^2)^2} \right\rangle.$$

   (a) Show that $\mathbf{V}$ is the velocity field of an incompressible potential flow in any region that does not contain $(0, 0)$.

   (b) Find a potential function for $\mathbf{V}$ and verify that it satisfies Laplace's equation.

   (c) Plot the vector field
$$\mathbf{W}(x, y) = \begin{cases} \mathbf{V}(x, y), & \text{if } x^2 + y^2 > 1 \\ (0, 0), & \text{otherwise} \end{cases}$$
   on the square region $-2 \leq x \leq 2$, $-2 \leq y \leq 2$. Note that for $x^2 + y^2 > 1$ this is the same vector field as in part (a).

   (d) Plot stream lines of the velocity field $\mathbf{V}(x, y)$ corresponding to the initial data
$$(x_0, y_0) = (-2..1), (-2, -.1), (-2, .5), (-2, -.5),$$
$$(-2, -1), (-2, 1)$$
   for $0 \leq t \leq 4$. Explain, in terms of the movement of fluid particles, what this plot might represent.

8.14 Reflecting Points

- Some related material is found in Section 11.7 of Stewart's **Multivariable Calculus: Concepts and Contexts**.

Suppose a curve C is given in the xy-plane and let P be a point on C. We will say that the point P is a reflecting point if there is another point Q on C so that the normal line to C through Q passes through P. If you imagine that the curve C is a made of mirror material, then a person standing at P could look at the point Q and see their reflection.

1. Find all of the reflecting points for the following curves:

    (a) $y = x^2$
    (b) $y = x^3$
    (c) $y = x^3 - x$

In a similar manner, if S is a surface in xyz-space and P is a point on S, then we say that the point P is a reflecting point if there is another point Q on S so that the normal line to S through Q passes through P.

2. Let $f(x, y) = x^2 + y^2$. Sketch the points (x, y) for which the point $(x, y, f(x, y))$ is a reflecting point for the surface given by the graph $z = f(x, y)$.

3. Repeat #2 for $f(x, y) = x^2 + 3y^2$.

4. Give an example of a surface in xyz-space which has no reflecting points.

5. Give an example of a surface in xyz-space for which every point is a reflecting point.

8.15 Volumes, Surfaces and Tangent Planes

- Related material is found in Sections 11.7, 12.3 and 12.5 of Stewart's **Multivariable Calculus: Concepts and Contexts**.

The graph of a function $z = f(x, y)$ is either everywhere concave up provided

$$f_{xx}(x, y) f_{yy}(x, y) - f_{xy}(x, y)^2 > 0 \text{ and } f_{xx}(x, y) > 0$$

for all (x, y).

8.15. VOLUMES, SURFACES AND TANGENT PLANES

1. Let $f(x,y) = x^2 + xy + 3y^2 + x - y$.

   (a) Show that this graph of this function is everywhere concave up.

   (b) Give a general formula for the tangent plane to $z = f(x,y)$ at the point $(a, b, f(a,b))$.

   (c) Compute the volume between the tangent plane and the graph for $-10 \leq x \leq 10$, $-10 \leq y \leq 10$ as an expression dependent upon (a, b).

   (d) Find the point at which the volume is a minimum. Be sure to apply the second derivative test to check your result.

2. Repeat #1 for the functions

   $$f(x,y) = x^4 + 2y^2 + e^{x-y},$$
   $$f(x,y) = x^2 + y^2 + 2x - y,$$
   $$f(x,y) = x^2 + y^2 + e^{x+y} - 3x.$$

3. Observe the location of the points with respect to the graphs of each of the functions above. What do you conjecture based upon your findings?

4. Prove your conjecture.

5. Is your conjecture dependent upon the choice of the set $-10 \leq x \leq 10$, $-10 \leq y \leq 10$? Check to see if the conjecture still holds if we change this set to be a rectangle, a circle or an ellipse.

Index

abs, 23, 62
acceleration, 65, 148
Agebra State, 3
Algebra State, 17
APPROX, 16
Approximate, 4
arc length, 56
area, 8
Author Expression, 1
author expression, 1
Autoscale Mode, 17
axes, 49, 58, 63

case sensitive, 8
center of mass, 104, 105, 108, 111, 113–115, 137
centripetal acceleration, 150
change of variables, 116, 117, 119, 120, 123, 125
circle of curvature, 149
component, 25
concave up, 174
conservative vector field, 128–130
continuity, 71, 73
contour plot, 40, 160, 164
COPROJECTION, 67
Coriolis acceleration, 150
critical point, 84, 86, 88, 158
CROSS, 28
curvature, 59, 66, 147
curve fitting, 155

cylindrical coordinates, 113

Declare, 3, 17
density, 104, 108, 110, 113, 114, 137
derivative, 52, 54, 77
DIF, 54, 77, 89
directional derivative, 82
divergence, 144
divergence free, 172
Divergence Theorem, 127, 143, 145
dot product, 24

e, 3
ENTER, 2
Exact, 4, 17
EXPAND, 9, 17
Expression, 1
expressions, 9

FACTOR, 10, 17
FACTOR (integers), 17
fixed point, 168
flow lines, 170
flux, 139, 141, 144, 145

GRAD, 80, 83
gradient, 82
gradient descent, 164
gradient field, 128, 129
Green's Theorem, 127, 135

INDEX

helix, 134, 152
Help, 11
Hessian matrix, 157, 160, 162

INT, 97, 99, 102
integral, 95, 97, 100, 103, 107, 108
irrotational, 171
ISOMETRIC, 49, 58, 63
ISOMETRICS, 67

Jacobian, 116, 117, 122, 123, 167

Lagrange multipliers, 84, 90, 94
Laplace's equation, 79
least squares, 155
level curve, 43
LIM, 73
limit, 71, 73
line, 29
line integral, 132, 134–136, 142, 145
linear approximation, 79, 80

mass, 104, 109, 110, 113–115, 137, 141
matrix, 20
maximum, 90, 92, 159
midpoint approximation, 95
minimum, 90, 92, 158, 164, 165
moment, 112
moments, 105

negative definite, 159
Newton's method, 162, 167, 168
NEWTONS, 88, 92, 93

Options, 17
osculating circle, 147

parametric plot, 15, 47
parametric surface, 67

partial derivative, 76, 80, 84
pi, 3
plane, 32
plot, 11
polar coordinates, 101–103, 143
positive definite, 158
potential, 128–131, 173
potential flow, 172
Precision Mode, 3, 16–18
principle unit normal, 148
projection, 26, 153

quadratic approximation, 160
quadratic fit, 156

rotation, 171

saddle, 160
scatter plot, 155
second derivative test, 160, 175
Set, 12
Simplification, 3
Simplification Options, 17
simplify, 10
SOLVE, 18, 30, 91, 103
solve, 88
space curve, 49, 57, 61
sphere, 67
spherical coordinates, 113, 138
SQRT, 7
Stewart, viii
Stokes' Theorem, 127, 142
stream lines, 171
SUB, 18, 22
SUM, 96
surface area, 106, 108
surface integral, 138, 139, 143
surface plot, 37

tangent plane, 79, 80

unit tangent vector, 58

VECTOR, 42, 48, 52, 55, 65, 98, 100, 104, 110
vector, 14, 19
vector field, 127–129, 131, 135, 144, 170
velocity, 65, 148
velocity field, 173
view plane, 153
volume, 98, 100, 103, 107, 109, 110, 113, 115, 175
vorticity, 171

work, 137